T0349830

The Global Nonlinear Stability of Minkowski Space for Self-Gravitating Massive Fields

Series in Applied and Computational Mathematics

ISSN: 2010-2739

Published

Vol. 1 Analysis of Singularities for Partial Differential Equations
by Shuxing Chen (Fudan University, China)

Vol. 2 The Hyperboloidal Foliation Method
by Philippe G. LeFloch (Université Pierre-et-Marie Curie, France) & Yue Ma (Université Pierre-et-Marie Curie, France)

Vol. 3 The Global Nonlinear Stability of Minkowski Space for Self-Gravitating Massive Fields
by Philippe G. LeFloch (Université Pierre-et-Marie Curie, France) & Yue Ma (Xi'an Jiaotong University, China)

Series in Applied and Computational Mathematics – Vol. 3

The Global Nonlinear Stability of Minkowski Space for Self-Gravitating Massive Fields

Philippe G LeFloch

Université Pierre-et-Marie Curie, France

Yue Ma

Xi'an Jiaotong University, China

NEW JERSEY • LONDON • SINGAPORE • BEIJING • SHANGHAI • HONG KONG • TAIPEI • CHENNAI

Published by

World Scientific Publishing Co. Pte. Ltd.
5 Toh Tuck Link, Singapore 596224
USA office: 27 Warren Street, Suite 401-402, Hackensack, NJ 07601
UK office: 57 Shelton Street, Covent Garden, London WC2H 9HE

Library of Congress Cataloging-in-Publication Data
Names: LeFloch, Philippe G., 1962– author. | Ma, Yue (Mathematician), author.
Title: The global nonlinear stability of Minkowski space for self-gravitating massive fields / by Philippe G. LeFloch (Université Pierre-et-Marie Curie, France), Yue Ma (Xi'an Jiaotong University, China).
Description: New Jersey : World Scientific, 2017. | Series: Series in applied and computational mathematics ; volume 3 | Includes bibliographical references.
Identifiers: LCCN 2017033398 | ISBN 9789813230859 (hardcover : alk. paper)
Subjects: LCSH: Gravitational fields. | Generalized spaces. | Scalar field theory. | Einstein field equations.
Classification: LCC QC178 .L3837 2017 | DDC 530.14--dc23
LC record available at https://lccn.loc.gov/2017033398

British Library Cataloguing-in-Publication Data
A catalogue record for this book is available from the British Library.

Printed in Singapore

The theory presented in this monograph establishes the first mathematically rigorous result on the *global nonlinear stability of self-gravitating matter* under small perturbations of an asymptotically flat, spacelike hypersurface of Minkowski spacetime. It allows one to exclude the existence of dynamically unstable, self-gravitating massive fields and, therefore, solves a long-standing open problem in General Relativity. By a significant extension of the *Hyperboloidal Foliation Method* they introduced in 2014, the authors establish global-in-time existence for the Einstein equations expressed as a *coupled wave-Klein-Gordon system* of partial differential equations. The metric and matter fields are sought for in Sobolev-type functional spaces, suitably defined from the translations and the boosts of Minkowski spacetime.

Preface

The theory presented in this Monograph establishes the first mathematically rigorous result about the *global nonlinear stability of self-gravitating matter* under small perturbations. It allows us to exclude the existence of dynamically unstable, self-gravitating massive fields and, therefore, solves a long-standing open problem in General Relativity.

We establish that Minkowski spacetime is nonlinearly stable in presence of a massive scalar field under suitable smallness conditions (for, otherwise, black holes might form). We formulate the initial value problem for the Einstein-massive scalar field equations, when the initial slice is a perturbation of an asymptotically flat, spacelike hypersurface in Minkowski space, and we prove that this perturbation disperses in future timelike directions so that the associated Cauchy development is future geodesically complete.

Our method of proof which we refer to as the *Hyperboloidal Foliation Method*, extends the standard 'vector field method' developed for massless fields and, importantly, does not use the scaling vector field of Minkowski space. We construct a foliation (of the interior of a light cone) by spacelike and asymptotically hyperboloidal hypersurfaces and we rely on a decomposition of the Einstein equations expressed in *wave gauge* and in a *semi-hyperboloidal frame*, in a sense defined in this Monograph. We focus here on the problem of the evolution of a spatially compact matter field, and we consider initial data coinciding, in a neighborhood of spacelike infinity, with a spacelike slice of Schwarzschild spacetime. We express the Einstein equations as a system of coupled nonlinear wave-Klein-Gordon equations (with differential constraints) posed on a curved space (whose metric is one of the unknowns).

The main challenge is to establish a global-in-time existence theory for *coupled wave-Klein-Gordon systems* in Sobolev-type spaces defined from the translations and the boosts of Minkowski spacetime, only. To this end, we rely on the following novel and robust techniques: new commutator estimates for hyperboloidal frames, sharp decay estimates for wave and Klein-Gordon equations, Sobolev and Hardy inequalities along the hyperboloidal foliation, quasi-null hyperboloidal structure of the Einstein equations, as well as integration arguments along characteristics and radial rays. Our proof also relies on an iterative procedure involving the components of the metric and the Klein-Gordon field, and on a hierarchy of low- and high-order energy estimates, which distinguishes between the metric components and between several levels of time dependency and regularity for the metric coefficients and the massive field.

Philippe LeFloch (Paris) and Yue Ma (Xi'an)

Contents

Preface vii

1. Introduction 1

   1.1 The nonlinear stability problem for the Einstein equations 1
   1.2 Statement of the main result 4

2. Overview of the Hyperboloidal Foliation Method 9

   2.1 The semi-hyperboloidal frame and the hyperboloidal frame 9
   2.2 Spacetime foliation and initial data set 11
   2.3 Coordinate formulation of the nonlinear stability property 14
   2.4 Bootstrap argument and construction of the initial data . 16
   2.5 Outline of the Monograph 17

3. Functional Analysis on Hyperboloids of Minkowski Spacetime 19

   3.1 Energy estimate on hyperboloids 19
   3.2 Sup-norm estimate based on curved characteristic integration 25
   3.3 Sup-norm estimate for wave equations with source 28
   3.4 Sup-norm estimate for Klein-Gordon equations 35
   3.5 Weighted Hardy inequality along the hyperboloidal foliation 36
   3.6 Sobolev inequality on hyperboloids 40
   3.7 Hardy inequality for hyperboloids 41
   3.8 Commutator estimates for admissible vector fields 43

4. Quasi-Null Structure of the Einstein-Massive Field System on Hyperboloids 45

   4.1 Einstein equations in wave coordinates 45

4.2 Analysis of the support 51
4.3 A classification of nonlinearities in the Einstein-massive field system 53
4.4 Estimates based on commutators and homogeneity 58
4.5 Basic structure of the quasi-null terms 59
4.6 Metric components in the semi-hyperboloidal frame 60
4.7 Wave gauge condition in the semi-hyperboloidal frame . . 62
4.8 Revisiting the structure of the quasi-null terms 65

5. Initialization of the Bootstrap Argument 69

5.1 The bootstrap assumption and the basic estimates 69
5.2 Estimates based on integration along radial rays 72

6. Direct Control of Nonlinearities in the Einstein Equations 73

6.1 L^∞ estimates 73
6.2 L^2 estimates 74

7. Direct Consequences of the Wave Gauge Condition 77

7.1 L^∞ estimates 77
7.2 L^2 estimates 81
7.3 Commutator estimates 84

8. Second-Order Derivatives of the Spacetime Metric 89

8.1 Preliminary 89
8.2 L^∞ estimates 91
8.3 L^2 estimates 92
8.4 Conclusion for general second-order derivatives 95
8.5 Commutator estimates 95

9. Sup-Norm Estimate Based on Characteristics 99

9.1 Main statement in this section 99
9.2 Application to quasi-null terms 103

10. Low-Order Refined Energy Estimate for the Spacetime Metric 105

10.1 Preliminary 105
10.2 Main estimate established in this section 107
10.3 Application of the refined energy estimate 110

11. Low-Order Refined Sup-Norm Estimate for the Metric and Scalar Field 113

11.1 Main estimates established in this section 113
11.2 First refinement on the metric components 114
11.3 First refinement for the scalar field 116
11.4 Second refinement for the scalar field and the metric . . . 118
11.5 A secondary bootstrap argument 120

12. High-Order Refined L^2 Estimates 125

12.1 Objective of this section and preliminary 125
12.2 Main estimates in this section 131
12.3 Applications to the derivation of refined decay estimates . 136

13. High-Order Refined Sup-Norm Estimates 139

13.1 Preliminary 139
13.2 Main estimate in this section 141

14. Low-Order Refined Energy Estimate for the Scalar Field 147

Appendix A Revisiting the wave-Klein-Gordon model 151

Appendix B Sup-norm estimate for the wave equations 153

Appendix C Sup-norm estimate for the Klein-Gordon equation 159

Appendix D Commutator estimates for the hyperboloidal frame 165

Bibliography 171

11. [illegible] Refined [illegible] Estimate for the [illegible]
and Scalar Fields [illegible]
11.1 [illegible]
11.2 [illegible]
11.3 [illegible] field [illegible]
11.4 [illegible] scalar field and the [illegible]
[illegible]

12. High [illegible]
[illegible]

Chapter 1

Introduction

1.1 The nonlinear stability problem for the Einstein equations

We consider Einstein's field equations of General Relativity for self-gravitating massive scalar fields and formulate the initial value problem when the initial data set is a perturbation of an asymptotically flat, space-like hypersurface in Minkowski spacetime. We then establish the existence of an Einstein development associated with this initial data set, which is proven to be an asymptotically flat and future geodesically complete spacetime. Recall that, in the case of vacuum spacetimes or massless scalar fields, such a nonlinear stability theory for Minkowski spacetime was first established by Christodoulou and Klainerman in their breakthrough work [12], which was later revisited by Lindblad and Rodnianski [46] via an alternative approach. Partial results on the global existence problem for the Einstein equations was obtained earlier by Friedrich [22; 23].

The *vacuum* Einstein equations are currently under particularly active development: this is illustrated by the recent contributions by Christodoulou [11] and Klainerman and Rodnianski [36] (on the formation of trapped surfaces) and by Klainerman, Rodnianski, and Szeftel [37] (on the L^2 curvature theorem). The Einstein equations coupled with massless fields such as the Maxwell field were also extensively studied; see for instance Bieri and Zipser [6] and Speck [54]; existence under slow decay conditions was established by Bieri [6].

The present Monograph offers a new method for the global analysis of the Einstein equations, which we refer to as the *Hyperboloidal Foliation Method* and allows us to investigate the global dynamics of massive fields and, especially, the *coupling between wave and Klein-Gordon equations.* This method was first outlined in [39; 41], together with references to earlier works, especially by Friedrich [22; 23], Klainerman [33], and Hörmander [27]. We hope that the present contribution will open a further direction of research concerning *matter spacetimes*, which need not be not Ricci-flat and may contain *massive fields.* In this direction, we refer to LeFloch et al. [5; 8; 25; 38; 43] for existence results on weakly regular matter spacetimes.

The *nonlinear stability problem for self-gravitating massive fields*, solved in the present Monograph[1], was a long-standing open problem for the past twenty five years since the publication of Christodoulou-Klainerman's book [12]. In the physics literature, blow-up mechanisms were proposed which suggest possible instabilities for self-gravitating massive fields. While the most recent numerical investigations [49] gave some confidence that Minkowski spacetime should be nonlinearly stable, the present work provides the *first mathematically rigorous proof that dynamically unstable solutions to the Einstein equations do not exist in presence of massive fields* (under suitable smallness conditions specified below). On the other hand, nonlinear stability would not hold when the mass is sufficiently large, since trapped surfaces and presumably black holes form from (large) perturbations of Minkowski spacetime [11].

Mathematically, the problem under consideration can be formulated (in the so-called wave gauge, see below) as a quasilinear system of *coupled nonlinear wave-Klein-Gordon equations,* supplemented with differential constraints and posed on a curved spacetime. The spacetime (Lorentzian) metric together with the scalar field defined on this spacetime are the unknowns of the Einstein-matter system. The Hyperboloidal Foliation Method introduced in this Monograph leads us to a *global-in-time theory* for this wave-Klein-Gordon system when initial data are provided on a spacelike hypersurface. Our proof is based on a substantial modification of the so-called vector field method, which have been applied to massless problems, only. Importantly, we do not use the scaling vector field of Minkowski spacetime, which is required to be able to handle Klein-Gordon equations.

In order to simplify the presentation of the method, in this Monograph we are interested in spatially compact matter fields and, therefore, we assume that the initial data coincide, in a neighborhood of spacelike infinity, with an asymptotically flat spacelike slice of Schwarzschild spacetime in wave coordinates. Our proof relies on several novel contributions: sharp

[1] We present here our method for a restricted class of initial data, while more general data as well as the theory of $f(R)$–modified gravity are treated in [42].

time-decay estimates for wave equations and Klein-Gordon equations on a curved spacetime, Sobolev and Hardy's inequalities on hyperboloids, quasi-null hyperboloidal structure of the Einstein equations and estimates based on integration along characteristics and radial rays. We also distinguish between low- and high-order energies for the metric coefficients and the massive field.

We refer to [39; 40; 41] for earlier work by the authors and to the companion work [42] for an extension to more general data and to the theory of modified gravity. We focus on $(3+1)$-dimensional problems since this is the dimension of main interest. As already mentioned, in the context of the Einstein equations, hyperboloidal foliations were introduced first by Friedrich [22; 23]. Of course, hyperboloidal foliations can be introduced in any number of dimensions, and should also lead to interesting results (see [47] in $(2+1)$ dimensions), but we do not pursue this here since the Einstein equations have rather different properties in these other dimensions.

For a different approach to the nonlinear stability of massive fields, we refer the reader to an ongoing research project by Q. Wang (outlined in ArXiv:1607.01466) which is aimed at generalizing Christodoulou-Klainerman's geometric method. An important recent development is provided by Fajman, Joudioux, and Smulevici [18; 19], who recently introduced a new vector field method based on a hyperboloidal foliation and aimed at dealing with global existence problems for massive kinetic equations; for this technique, we also refer to Smulevici [53]. Hyperboloidal foliations are also useful to analyze the blow-up of solutions for, for instance, focusing wave equations, as investigated by Burtscher and Donninger [7].

Furthermore, we also recall that nonlinear wave equations of Klein-Gordon-type posed on possibly curved spacetimes have been the subject of extensive research in the past two decades, and we will not try to review this vast literature and we refer the interested reader to, for instance, Bachelot [3; 4], Delort et al. [16; 17], Katayama [30; 31], and Shatah [51; 52], as well as Germain [24] and Ionescu and Pasauder [29]; see also [27; 28; 55] and the references cited therein. Importantly, the use of hyperboloidal foliations leads to robust and efficient numerical methods, as demonstrated by a variety of approaches by Ansorg and Macedo [1], Frauendiener [20; 21], Hilditch et al. [26; 56], Moncrief and Rinne [48], Rinne [50], and Zenginoglu [57; 58].

1.2 Statement of the main result

We thus consider the *Einstein equations* for an unknown spacetime (M, g), that is,

$$G_{\alpha\beta} := R_{\alpha\beta} - \frac{R}{2} g_{\alpha\beta} = 8\pi\, T_{\alpha\beta}, \tag{1.1}$$

where $R_{\alpha\beta}$ denotes the Ricci curvature of (M, g), $R = g^{\alpha\beta} R_{\alpha\beta}$ its scalar curvature, and $G_{\alpha\beta}$ is referred to as the Einstein tensor. Our main unknown in (1.1) is a Lorentzian metric $g_{\alpha\beta}$ defined on a topological 4-manifold M. By convention, Greek indices $\alpha, \beta, \dots$ take values $0, 1, 2, 3$, while Latin indices $i, j, \dots$ takes values $1, 2, 3$ (as, for instance, in (1.5) below). In this work, we are interested in non-vacuum spacetimes when the matter content is described by a massive scalar field denoted by $\phi : M \to \mathbb{R}$ with potential $V = V(\phi)$. The stress-energy tensor of such a field reads

$$T_{\alpha\beta} := \nabla_\alpha \phi \nabla_\beta \phi - \Big(\frac{1}{2} \nabla_\gamma \phi \nabla^\gamma \phi + V(\phi) \Big) g_{\alpha\beta}. \tag{1.2}$$

Recall that from the contracted Bianchi identities $\nabla^\alpha G_{\alpha\beta} = 0$, we can derive an evolution equation for the scalar field and, in turn, formulate the Einstein–massive field system as the system of quasilinear partial differential equations (in any choice of coordinates at this stage)

$$R_{\alpha\beta} = 8\pi \big(\nabla_\alpha \phi \nabla_\beta \phi + V(\phi)\, g_{\alpha\beta} \big), \tag{1.3a}$$

$$\Box_g \phi - V'(\phi) = 0. \tag{1.3b}$$

Without loss of generality, throughout we assume that the potential is quadratic in ϕ, i.e.

$$V(\phi) = \frac{c^2}{2} \phi^2, \tag{1.4}$$

where $c^2 > 0$ is referred to as the mass density of the scalar field. The equation (1.3b) is nothing but a Klein-Gordon equation posed on an (unknown) curved spacetime.

The Cauchy problem for the Einstein equations can be formulated as follows; cf., for instance, Choquet-Bruhat's textbook [9]. First of all, let us recall that an *initial data set* for the Einstein equations consists of a Riemannian 3-manifold $(\overline{M}, \overline{g})$, a symmetric 2-tensor field K defined on $\overline{M}$, and two scalar fields ϕ_0 and ϕ_1 also defined on $\overline{M}$. A *Cauchy development of the initial data set* $(\overline{M}, \overline{g}, K, \phi_0, \phi_1)$, by definition, is a $(3+1)$-dimensional Lorentzian manifold (M, g) satisfying the following two properties:

- There exists an embedding $i : \overline{M} \to M$ such that the (pull-back) induced metric $i^*(g) = \overline{g}$ coincides with the prescribed metric $\overline{g}$, while the second fundamental form of $i(\overline{M}) \subset M$ coincides with the prescribed 2-tensor K. In addition, by denoting by n the (future-oriented) unit normal to $i(\overline{M})$, the restriction (to the hypersurface $i(\overline{M})$) of the field ϕ and its Lie derivative $\mathcal{L}_n\phi$ coincides with the data ϕ_0 and ϕ_1 respectively.
- The manifold (M, g) satisfies the Einstein equations (1.3a) and, consequently, the scalar field ϕ satisfies the Klein-Gordon equation (1.3b).

As is well-known, in order to fulfill the equations (1.3a), the initial data set cannot be arbitrary but must satisfy Einstein's constraint equations:

$$\overline{R} - K_{ij} K^{ij} + (K_i^i)^2 = 8\pi T_{00}, \qquad \overline{\nabla}^i K_{ij} - \overline{\nabla}_j K_l^l = 8\pi T_{0j}, \tag{1.5}$$

where $\overline{R}$ and $\overline{\nabla}$ are the scalar curvature and Levi-Civita connection of the manifold $(\overline{M}, \overline{g})$, respectively, while the mass-energy density T_{00} and the momentum vector T_{0i} are determined from the data ϕ_0, ϕ_1 (in view of the expression (1.2) of the stress-energy tensor).

Our main result established in the present Monograph can be stated as follows.

Theorem 1.1. (Nonlinear stability of Minkowski spacetime for self-gravitating massive fields. Geometric version). *Consider the Einstein-massive field system* (1.3) *when the initial data set* $(\overline{M}, \overline{g}, K, \phi_0, \phi_1)$ *satisfies Einstein's constraint equations* (1.5) *and is close to an asymptotically flat slice of the (vacuum) Minkowski spacetime and, more precisely, coincides in a neighborhood of spacelike infinity with a spacelike slice of a Schwarzschild spacetime with sufficiently small ADM mass. The corresponding initial value problem admits a globally hyperbolic Cauchy development, which represents an asymptotically flat and future geodesically complete spacetime.*

We observe that the existence of initial data sets satisfying the conditions above was established by Corvino and Schoen [15]; see also Chrusciel and Delay [14] and the recent review [13]. Although the main focus therein is on vacuum spacetimes, it is straightforward to include matter fields by observing[2] that classical existence theorems [9] provide the existence of

[2]The authors thank J. Corvino for pointing this out to them.

non-trivial initial data in the "interior region" and that Corvino-Schoen's glueing construction is purely local in space.

We are going to formulate the Einstein-massive field system as coupled partial differential equations. This is achieved by introducing *wave coordinates* denoted by x^α, satisfying the wave equation $\Box_g x^\alpha = 0$ $(\alpha = 0, \dots, 3)$. From (1.3), we will see that, in wave coordinates, the Ricci curvature operator reduces to the wave operator on the metric coefficients and, in fact, (cf. Lemma 4.1, below)

$$\widetilde{\Box}_g h_{\alpha\beta} = F_{\alpha\beta}(h; \partial h, \partial h) - 16\pi \partial_\alpha \phi \partial_\beta \phi - 16\pi V(\phi) g_{\alpha\beta}, \tag{1.6a}$$

$$\widetilde{\Box}_g \phi - V'(\phi) = 0, \tag{1.6b}$$

where $\widetilde{\Box}_g := g^{\alpha\beta}\partial_\alpha\partial_\beta$ is referred to as the *reduced wave operator*, and $h_{\alpha\beta} := g_{\alpha\beta} - m_{\alpha\beta}$ denotes the curved part of the unknown metric. The nonlinear terms $F_{\alpha\beta}(h; \partial h, \partial h)$ are quadratic in first-order derivatives of the metric. Of course, that the system (1.6) must be supplemented with Einstein's constraints (1.5) as well as the wave gauge conditions $\Box_g x^\alpha = 0$, which both are first-order differential constraints on the metric.

In order to establish a global-in-time existence theory for the above system, several major challenges are overcome in the present work:

- Most importantly, we cannot use the scaling vector field $S := r\partial_r + t\partial_t$, since the Klein-Gordon equation is not kept conformally invariant by this vector field.
- In addition to null terms which are standard in the theory of quasilinear wave equations, in the nonlinearity $F_{\alpha\beta}(h; \partial h, \partial h)$ we must also handle *quasi null terms*, as we call them, which will be controlled by relying on the wave gauge condition.
- The structure of the nonlinearities in the Einstein equations must be carefully studied in order to exclude instabilities that may be induced by the *massive scalar field*.

In addition to the refined estimates on the commutators for hyperboloidal frames[3] and the sharp L^∞-L^∞ estimates for wave equations and Klein-Gordon equations already introduced by the authors in the first part [41], we need the following new arguments of proof (further discussed below):

[3] A fundamental observation in [39] is that commutators of hyperboloidal frames and Lorentz boosts enjoy good properties, which had not been observed in earlier works on the subject. In the notation presented below, this is especially true of $[L_a, (s/t)\partial_t]$ and $[L_a, \overline{\partial}_b]$.

- Formulation of the Einstein equations in wave gauge in the semi-hyperboloidal frame.
- Energy estimates at arbitrary order on a background Schwarzschild space in wave gauge.
- Refined estimates for nonlinear wave equations, that are established by integration along characteristics or radial rays.
- Estimates of quasi-null terms in wave gauge, for which we rely on, both, the tensorial structure of the Einstein equations and the wave gauge condition.
- New weighted Hardy inequality along the hyperboloidal foliation.

A precise outline of the content of this Monograph will be given at the end of the following section, after introducing further notation.

Acknowledgements

The results established in this Monograph were presented in July 2015 at the Mini-Symposium on "Mathematical Problems of General Relativity" (EquaDiff Conference), organized by S. Klainerman and J. Szeftel at the University of Lyon (France). They also led to a graduate course [38] taught by the first author at the Institute Henri Poincaré (Paris) during the quarter program "Mathematical General Relativity: the 100th anniversary" organized by L. Andersson, S. Klainerman, and P.G. LeFloch in the Fall 2015. The first author (PLF) gratefully acknowledges support from the Simons Center for Geometry and Physics, Stony Brook University, during the one-month Program "Mathematical Method in General Relativity" hold in January 2015 and organized by M. Anderson, S. Klainerman, P.G. LeFloch, and J. Speck.

Chapter 2

Overview of the Hyperboloidal Foliation Method

2.1 The semi-hyperboloidal frame and the hyperboloidal frame

Consider the $(3+1)$-dimensional Minkowski spacetime with signature $(-,+,+,+)$. In Cartesian coordinates, we write $(t,x) = (x^0, x^1, x^2, x^3)$ with $r^2 := |x|^2 = (x^1)^2 + (x^2)^2 + (x^3)^2$, and we use the partial derivative fields ∂_0 and ∂_a, as well as the Lorentz boosts $L_a := x^a\partial_t + t\partial_a$ and their "normalized" version $\frac{L_a}{t} = \frac{x^a}{t}\partial_t + \partial_a$. We primarily deal with functions defined in the interior of the future light cone from the point $(1,0,0,0)$, denoted by

$$\mathcal{K} := \{(t,x)\,/\,r < t-1\}.$$

To foliate this domain, we consider the hyperboloidal hypersurfaces with hyperbolic radius $s > 0$, defined by

$$\mathcal{H}_s := \big\{(t,x)\,/\,t^2 - r^2 = s^2; \quad t > 0\big\}$$

with $s \geq 1$. In particular, we can introduce the following subset of $\mathcal{K}$ limited by two hyperboloids (with $s_0 < s_1$)

$$\mathcal{K}_{[s_0,s_1]} := \big\{(t,x)\,/\,s_0^2 \leq t^2 - r^2 \leq s_1^2; \quad r < t-1\big\}$$

whose boundary contains a section of the light cone $\mathcal{K}$.

With these notations, the *semi-hyperboloidal frame* is, by definition,

$$\underline{\partial}_0 := \partial_t, \qquad \underline{\partial}_a := \frac{x^a}{t}\partial_t + \partial_a, \qquad a = 1,2,3. \tag{2.1}$$

Note that the three vectors $\underline{\partial}_a$ generate the tangent space to the hyperboloids. For some of our statements (for instance in Proposition 3.5), it will be convenient to also use the vector field $\underline{\partial}_\perp := \partial_t + \frac{x^a}{t}\partial_a$, which is orthogonal to the hyperboloids (and is proportional to the scaling vector field).

Furthermore, given a multi-index $I = (\alpha_n, \alpha_{n-1}, \ldots, \alpha_1)$ with $\alpha_i \in \{0,1,2,3\}$, we use the notation $\partial^I := \partial_{\alpha_n}\partial_{\alpha_{n-1}}\cdots\partial_{\alpha_1}$ for the product of n partial derivatives and, similarly, for $J = (a_n, a_{n-1}, \ldots, a_1)$ with $a_i \in \{1,2,3\}$ we write $L^J = L_{a_n}L_{a_{n-1}}\cdots L_{a_1}$ for the product of n Lorentz boosts.

Associated with the semi-hyperboloidal frame, one has the dual frame $\underline{\theta}^0 := dt - \frac{x^a}{t}\,dx^a$, $\underline{\theta}^a := dx^a$. The (dual) semi-hyperboloidal frame and the (dual) natural Cartesian frame are related via

$$\underline{\partial}_\alpha = \Phi_\alpha^{\alpha'}\partial_{\alpha'}, \quad \partial_\alpha = \Psi_\alpha^{\alpha'}\underline{\partial}_{\alpha'}, \qquad \underline{\theta}^\alpha = \Psi_{\alpha'}^\alpha\,dx^{\alpha'}, \quad dx^\alpha = \Phi_{\beta'}^\alpha\underline{\theta}^{\alpha'},$$

in which the transition matrix (Φ_α^β) and its inverse (Ψ_α^β) are

$$(\Phi_\alpha^\beta) = \begin{pmatrix} 1 & 0 & 0 & 0 \\ x^1/t & 1 & 0 & 0 \\ x^2/t & 0 & 1 & 0 \\ x^3/t & 0 & 0 & 1 \end{pmatrix}, \qquad (\Psi_\alpha^\beta) = \begin{pmatrix} 1 & 0 & 0 & 0 \\ -x^1/t & 1 & 0 & 0 \\ -x^2/t & 0 & 1 & 0 \\ -x^3/t & 0 & 0 & 1 \end{pmatrix}.$$

With this notation, for any two-tensor $T_{\alpha\beta}\,dx^\alpha \otimes dx^\beta = \underline{T}_{\alpha\beta}\underline{\theta}^\alpha \otimes \underline{\theta}^\beta$, we can write $\underline{T}_{\alpha\beta} = T_{\alpha'\beta'}\Phi_\alpha^{\alpha'}\Phi_\beta^{\beta'}$ and $T_{\alpha\beta} = \underline{T}_{\alpha'\beta'}\Psi_\alpha^{\alpha'}\Psi_\beta^{\beta'}$. We also have the similar decompositions $\underline{T}^{\alpha\beta} = T^{\alpha'\beta'}\Phi_{\alpha'}^\alpha\Phi_{\beta'}^\beta$ and $T^{\alpha\beta} = \underline{T}^{\alpha'\beta'}\Psi_{\alpha'}^\alpha\Psi_{\beta'}^\beta$.

Lemma 2.1 (Decomposition of the wave operator).
For every smooth function u defined in the future light-cone $\mathcal{K}$, the flat wave operator in the semi-hyperboloidal frame reads

$$\Box u = -\frac{s^2}{t^2}\partial_t\partial_t u - \frac{3}{t}\partial_t u - \frac{x^a}{t}\left(\partial_t\underline{\partial}_a u + \underline{\partial}_a\partial_t u\right) + \sum_a \underline{\partial}_a\underline{\partial}_a u. \tag{2.2}$$

Within the future cone $\mathcal{K}$, we introduce the change of variables $\overline{x}^0 = s := \sqrt{t^2 - r^2}$ and $\overline{x}^a = x^a$ and the associated frame which we refer to as the *hyperboloidal frame:*

$$\begin{aligned} \overline{\partial}_0 &:= \partial_s = \frac{s}{t}\partial_t = \frac{\overline{x}^0}{t}\partial_t = \frac{\sqrt{t^2-r^2}}{t}\partial_t, \\ \overline{\partial}_a &:= \partial_{\overline{x}^a} = \frac{\overline{x}^a}{t}\partial_t + \partial_a = \frac{x^a}{t}\partial_t + \partial_a. \end{aligned} \tag{2.3}$$

The transition matrices between the hyperboloidal frame and the Cartesian frame read

$$(\overline{\Phi}_\alpha^\beta) = \begin{pmatrix} s/t & 0 & 0 & 0 \\ x^1/t & 1 & 0 & 0 \\ x^2/t & 0 & 1 & 0 \\ x^3/t & 0 & 0 & 1 \end{pmatrix}, \qquad (\overline{\Psi}_\alpha^{\ \beta}) := (\overline{\Phi}_\alpha^\beta)^{-1} = \begin{pmatrix} t/s & 0 & 0 & 0 \\ -x^1/s & 1 & 0 & 0 \\ -x^2/s & 0 & 1 & 0 \\ -x^3/s & 0 & 0 & 1 \end{pmatrix},$$

so that $\overline{\partial}_\alpha = \overline{\Phi}_\alpha^\beta \partial_\beta$ and $\partial_\alpha = \overline{\Psi}_\alpha^\beta \overline{\partial}_\beta$. Observe also that the dual hyperboloidal frame is $d\overline{x}^0 := ds = \frac{t}{s}\,dt - \frac{x^a}{s}\,dx^a$ and $d\overline{x}^a := dx^a$, while the Minkowski metric in the hyperboloidal frame reads

$$(\overline{m}^{\alpha\beta}) = \begin{pmatrix} -1 & -x^1/s & -x^2/s & -x^3/s \\ -x^1/s & 1 & 0 & 0 \\ -x^2/s & 0 & 1 & 0 \\ -x^3/s & 0 & 0 & 1 \end{pmatrix}.$$

A given tensor can be expressed in any of the above three frames: the standard frame $\{\partial_\alpha\}$, the semi-hyperboloidal frame $\{\underline{\partial}_\alpha\}$, and the hyperboloidal frame $\{\overline{\partial}_\alpha\}$. We use Roman letters, underlined Roman letters and overlined Roman letters for the corresponding components of a tensor expressed in different frame. For example, $T^{\alpha\beta}\partial_\alpha \otimes \partial_\beta$ also reads $T^{\alpha\beta}\partial_\alpha \otimes \partial_\beta = \underline{T}^{\alpha\beta}\underline{\partial}_\alpha \otimes \underline{\partial}_\beta = \overline{T}^{\alpha\beta}\overline{\partial}_\alpha \otimes \overline{\partial}_\beta$, where $\overline{T}^{\alpha\beta} = \overline{\Psi}_{\alpha'}^\alpha \overline{\Psi}_{\beta'}^\beta T^{\alpha'\beta'}$ and, moreover, by setting $M := \max_{\alpha\beta}|T^{\alpha\beta}|$, in the hyperboloidal frame we have the uniform bounds[1] $(s/t)^2\,|\overline{T}^{00}| + (s/t)\,|\overline{T}^{a0}| + |\overline{T}^{ab}| \lesssim M$.

2.2 Spacetime foliation and initial data set

We now discuss the construction of the initial data by following the notation in [9, Sections VI.2 and VI.3]. We are interested in a time-oriented spacetime (M, g) that is endowed with a Lorentzian metric g with signature $(-,+,+,+)$ and admits a global foliation by spacelike hypersurfaces $M_t \simeq \{t\} \times \mathbb{R}^3$. The foliation is determined by a time function $t : M \to [0, +\infty)$. We introduce local coordinates adapted to the above product structure, that is, $(x^\alpha) = (x^0 = t, x^i)$, and we choose the basis of vectors (∂_i) as the 'natural frame' of each slice M_t, and this also defines the 'natural frame' (∂_t, ∂_i) on the spacetime M. By definition, the 'Cauchy adapted frame' is $e_i = \partial_i$ and $e_0 = \partial_t - \beta^i\partial_i$, where $\beta = \beta^i\partial_i$ is a time-dependent field, tangent to M_t and is called the *shift vector*, and we impose the restriction that e_0 is orthogonal to each hypersurface M_t. The dual frame (θ^α) of the Cauchy adapted frame (e_α), by definition, is $\theta^0 := dt$ and $\theta^i := dx^i + \beta^i dt$ and the spacetime metric reads

$$g = -N^2\theta^0\theta^0 + g_{ij}\theta^i\theta^j, \tag{2.4}$$

where the function $N > 0$ is referred to as the *lapse function* of the foliation.

[1] Here and in the rest of this paper, the notation $A \lesssim B$ is used when $A \leq CB$ and C is already known to be bounded (at the stage of the analysis).

We denote by $\overline{g} = \overline{g}_t$ the induced Riemannian metric associated with the slices M_t and by $\overline{\nabla}$ the Levi-Civita connection of $\overline{g}$. We also introduce the *second fundamental form* $K = K_t$ defined by

$$K(X,Y) := -g(\nabla_X n, Y)$$

for all vectors X, Y tangent to the slices M_t, where n denotes the future-oriented, unit normal to the slices. In the Cauchy adapted frame, it reads

$$K_{ij} = -\frac{1}{2N}\Big(\langle e_0, g_{ij}\rangle - g_{lj}\partial_i\beta^l - g_{il}\partial_j\beta^l\Big).$$

Here, we use the notation $\langle e_0, g_{ij}\rangle$ for the action of the vector field e_0 on the function g_{ij}. Next, we define the *time-operator* D_0 acting on a two-tensor defined on the slice M_t by $D_0 T_{ij} = \langle e_0, T_{ij}\rangle - T_{lj}\partial_i\beta^l - T_{il}\partial_j\beta^l$, which is again a two-tensor on M_t. With this notation, we have

$$K = -\frac{1}{2N} D_0\overline{g}.$$

In order to express the field equations (1.3) as a system of partial differential equations (PDE) in wave coordinates, we need first to turn the geometric initial data set $(\overline{M}, \overline{g}, K, \phi_0, \phi_1)$ into a "PDE initial data set". Since the equations are second-order, we need to know the data $g_{\alpha\beta}|_{\{t=2\}} = g_{0,\alpha\beta}$, $\partial_t g_{\alpha\beta}|_{\{t=2\}} = g_{1,\alpha\beta}$, $\phi|_{\{t=2\}} = \phi_0$, $\partial_t\phi|_{\{t=2\}} = \phi_1$, that is, the metric and the scalar field and their time derivative evaluated on the initial hypersurface $\{t = 2\}$. We claim that these data can be precisely determined from the prescribed geometric data $(\overline{g}, K, \phi_0, \phi_1)$, as follows. The PDE initial data satisfy:

- 4 Gauss-Codazzi equations which form the system of Einstein's constraints, and
- 4 equations deduced from the (restriction of the) wave gauge condition.

For the PDE initial data we have to determine 22 components, and the geometric initial data provide us with $(\overline{g}_{ab}, K_{ab}, \phi_0, \phi_1)$, that is, 14 components in total. The remaining degrees of freedom are exactly determined by the above 8 equations. The well-posedness of the system composed by the above 8 equations is a trivial property. In this work, we are concerned with the evolution part of the Einstein equations and our discussion is naturally based directly on the PDE initial data set.

The initial data sets considered in the present article are taken to be "near" initial data sets generating the Minkowski metric (i.e. without matter field). More precisely, we consider initial data sets which coincide, outside a spatially compact set $\{|x| \leq 1\}$, with an asymptotically flat, spacelike hypersurface in a Schwarzschild spacetime with sufficiently small ADM mass. The following observation is in order. The main challenge overcome by the hyperboloidal foliation method applied to (1.6) concerns the part of the solution supported in the region $\mathcal{K}_{[2,+\infty)}$ or, more precisely, the global evolution of initial data posed on an asymptotically hyperbolic hypersurface. (See [42] for further details.) To guarantee this, the initial data posed on the hypersurface $\{t = 2\}$ should have its support contained in the unit ball $\{r < 1\}$. Of course, in view of the positive mass theorem (associated with the constraint equation (1.5)), admissible non-trivial initial data must have a non-trivial tail at spatial infinity, that is,

$$m_S := \lim_{r\to+\infty} \int_{\Sigma_r} (\partial_j g_{ij} - \partial_i g_{jj}) n^i d\Sigma, \tag{2.5}$$

where n is the outward unit norm to the sphere Σ_r with radius r. Therefore, an initial data (unless it identically vanishes) cannot be supported in a compact region.

To bypass this difficulty, we make the following observation: first, the Schwarzschild spacetime provides us with an exact solution to (1.3), that is, the equations (1.6) (when expressed with wave coordinates). So, we assume that our initial data g_0 and g_1 coincide with the restriction of the Schwarzschild metric and its time derivative, respectively (again in wave coordinates) on the initial hypersurface $\{t = 2\}$ outside the unit ball $\{r < 1\}$. Outside the region $\mathcal{K}_{[2,+\infty)}$, we prove that the solution coincides with Schwarzschild spacetime and the global existence problem can be posed in the region $\mathcal{K}_{[2,+\infty)}$.

We can also formulate the Cauchy problem directly with initial data posed on a hyperboloidal hypersurface. This appears to be, both, geometrically and physically natural. As we demonstrated earlier in [39], the analysis of nonlinear wave equations is also more natural in such a setup and may lead us to *uniform bounds* for the energy of the solutions. Yet another approach would be to pose the Cauchy problem on a light cone, but while it is physically appealing, such a formulation would introduce spurious technical difficulties (i.e. the regularity at the tip of the cone) and does not appear to be very convenient from the analysis viewpoint.

The Schwarzschild metric in standard wave coordinates (x^0, x^1, x^2, x^3) takes the form (cf. [2]):

$$g_{S00} = -\frac{r - m_S}{r + m_S}, \quad g_{Sab} = \frac{r + m_S}{r - m_S}\omega_a\omega_b + \frac{(r + m_S)^2}{r^2}(\delta_{ab} - \omega_a\omega_b) \tag{2.6}$$

with $\omega_a := x_a/r$. Furthermore, in order to distinguish between the behavior in the small and in the large, we introduce a smooth cut-off function $\chi : \mathbb{R}^+ \to \mathbb{R}$ (fixed once for all) satisfying $\chi(\tau) = 0$ for $\tau \in [0, 1/3]$ while $\chi(\tau) = 1$ for $\tau \in [2/3, +\infty)$.

Definition 2.1. An initial data set for the Einstein-massive field system posed on the initial hypersurface $\{t = 2\}$ is said to be a spatially compact perturbation of Schwarzschild spacetime or a compact Schwarzschild perturbation, in short, if outside a compact set it coincides with the (vacuum) Schwarzschild space.

The proof of the following result is postponed to Chapter 4.2, after investigating the nonlinear structure of the Einstein-massive field system.

Proposition 2.1. *Let $(g_{\alpha\beta}, \phi)$ be a solution to the system (1.6) whose initial data is a compact Schwarzschild perturbation, then $(g_{\alpha\beta} - g_{S\alpha\beta})$ is supported in the region $\mathcal{K}$ and vanishes in a neighborhood of the boundary $\partial_B\mathcal{K} := \{r = t - 1, t \geq 2\}$.*

2.3 Coordinate formulation of the nonlinear stability property

We introduce the restriction

$$\mathcal{H}_s^* := \mathcal{H}_s \cap \mathcal{K}$$

of the hyperboloid to the light cone and we consider the energy functionals

$$E_{g,c^2}(s,u) := \int_{\mathcal{H}_s} \Big(- g^{00}|\partial_t u|^2 + g^{ab}\partial_a u \partial_b u + \sum_a \frac{2x^a}{t} g^{a\beta}\partial_\beta u \partial_t u + c^2 u^2 \Big)\, dx,$$

$$E^*_{g,c^2}(s,u) := \int_{\mathcal{H}^*_s} \Big(- g^{00}|\partial_t u|^2 + g^{ab}\partial_a u \partial_b u + \sum_a \frac{2x^a}{t} g^{a\beta}\partial_\beta u \partial_t u + c^2 u^2 \Big)\, dx,$$

and, for the flat Minkowski background,

$$E_{M,c^2}(s,u) := \int_{\mathcal{H}_s} \Big(|\partial_t u|^2 + \sum_a |\partial_a u|^2 + \sum_a \frac{2x^a}{t} \partial_a u \partial_t u + c^2 u^2 \Big)\, dx,$$

$$E^*_{M,c^2}(s,u) := \int_{\mathcal{H}^*_s} \Big(|\partial_t u|^2 + \sum_a |\partial_a u|^2 + \sum_a \frac{2x^a}{t} \partial_a u \partial_t u + c^2 u^2 \Big)\, dx.$$

We have the alternative form

$$\begin{aligned} E_{M,c^2}(s,u) &= \int_{\mathcal{H}_s} \Big((s/t)^2|\partial_t u|^2 + \sum_a |\underline{\partial}_a u|^2 + c^2 u^2 \Big)\, dx \\ &= \int_{\mathcal{H}_s} \Big(|\partial_t u + (x^a/t)\partial_a u|^2 + \sum_{a<b} |t^{-1}\Omega_{ab} u|^2 + c^2 u^2 \Big)\, dx, \end{aligned}$$

where $\Omega_{ab} := x^a\partial_b - x^b\partial_a$ denotes the spatial rotations. When the parameter c is taken to vanish, we also use the short-hand notation $E_g^*(s,u) := E_{g,0}^*(s,u)$ and $E_g(s,u) := E_{g,0}(s,u)$. In addition, for all $p \in [1,+\infty)$, the L^p norms on the hyperboloids endowed with the (flat) measure dx are denoted by

$$\|u\|_{L^p_f(\mathcal{H}_s)}^p := \int_{\mathcal{H}_s} |u|^p dx = \int_{\mathbb{R}^3} \big|u(\sqrt{s^2+r^2},x)\big|^p dx$$

and the L^P norms on the interior of $\mathcal{H}_s$ by

$$\|u\|_{L^p(\mathcal{H}_s^*)}^p := \int_{\mathcal{H}_s\cap\mathcal{K}} |u|^p dx = \int_{r\leq(s^2-1)/2} \big|u(\sqrt{s^2+r^2},x)\big|^p dx.$$

We are now in a position to state our main result for the Einstein system (1.6). The principal part of our system is the reduced wave operator associated with the curved metric g and we can write the decomposition

$$\widetilde{\Box}_g = g^{\alpha\beta}\partial_\alpha\partial_\beta = \Box + H^{\alpha\beta}\partial_\alpha\partial_\beta, \tag{2.7}$$

in which $H^{\alpha\beta} := m^{\alpha\beta} - g^{\alpha\beta}$ are functions of $h = (h_{\alpha\beta})$. When h is sufficiently small, $H^{\alpha\beta}(h)$ can be expressed as a power series in the components $h_{\alpha\beta}$ and vanishes at first-order at the origin. Our analysis will (only) use the translation and boost Killing fields associated with the flat wave operator $\Box$ in the coordinates under consideration.

Theorem 2.1. (Nonlinear stability of Minkowski spacetime for self-gravitating massive fields. Formulation in coordinates.) *Consider the Einstein-massive field equations* (1.6) *together with an initial data set satisfying the constraints and prescribed on the hypersurface* $\{t=2\}$*:*

$$\begin{aligned} g_{\alpha\beta}|_{\{t=2\}} &= g_{0,\alpha\beta}, & \partial_t g_{\alpha\beta}|_{\{t=2\}} &= g_{1,\alpha\beta}, \\ \phi|_{\{t=2\}} &= \phi_0, & \partial_t\phi|_{\{t=2\}} &= \phi_1, \end{aligned} \tag{2.8}$$

which, on $\{t=2\}$ *outside the unit ball* $\{r<1\}$*, is assumed to coincide with the restriction of Schwarzschild spacetime of mass* m_S *(in the wave gauge* (2.6)*), i.e. in* $\{r=|x|\geq 1\}$

$$g_{\alpha\beta}(2,\cdot) = g_{S\alpha\beta}, \qquad \partial_t g_{\alpha\beta}(2,\cdot) = \phi(2,\cdot) = \partial_t\phi(2,\cdot) = 0.$$

Then, for any sufficiently large integer N*, there exist constants* $\varepsilon_0, C_1, \delta > 0$ *and such that provided* $m_S \leq \varepsilon \leq \varepsilon_0$ *and*

$$\sum_{\alpha,\beta} \|\partial g_{0,\alpha\beta}, g_{1,\alpha\beta}\|_{H^N(\{r<1\})} + \|\phi_0\|_{H^{N+1}(\{r<1\})} + \|\phi_1\|_{H^N(\{r<1\})} \leq \epsilon \tag{2.9}$$

hold at the initial time, then the solution associated with the initial data (2.8) *exists for all times* $t\geq 2$ *and, furthermore,*

$$\begin{aligned} &E_M(s,\partial^I L^J h_{\alpha\beta})^{1/2} \leq C_1\varepsilon s^\delta, && |I|+|J|\leq N, \\ &E_{M,c^2}(s,\partial^I L^J \phi)^{1/2} \leq C_1\varepsilon s^{\delta+1/2}, && |I|+|J|\leq N, \\ &E_{M,c^2}(s,\partial^I L^J \phi)^{1/2} \leq C_1\varepsilon s^\delta, && |I|+|J|\leq N-4. \end{aligned} \tag{2.10}$$

2.4 Bootstrap argument and construction of the initial data

We will rely on a bootstrap argument, which can be sketched as follows. We begin with our main system (1.6) supplemented with initial data on the initial hyperboloid $\mathcal{H}_2$, that is, $g_{\alpha\beta}|_{\mathcal{H}_2}$, $\partial_t g_{\alpha\beta}|_{\mathcal{H}_2}$, $\phi|_{\mathcal{H}_2}$, and $\partial_t\phi|_{\mathcal{H}_2}$. First of all, since the initial data is posed on $\{t=2\}$ and is sufficiently small, we need first to construct its restriction on the initial hyperboloid $\mathcal{H}_2$. Since the data are compactly supported, this is immediate by the standard local existence theorem (see [39, Chap. 11] for the details). We also observe that when the initial data posed on $\{t=2\}$ are sufficiently small, i.e. (2.9) holds, then the corresponding data on $\mathcal{H}_2$ satisfies the bounds

$$
\begin{aligned}
\|\partial_a\partial^I L^J h_{\alpha\beta}\|_{L^2(\mathcal{H}_2^*)} + \|\partial_t\partial^I L^J h_{\alpha\beta}\|_{L^2(\mathcal{H}_2^*)} &\leq C_0\,\varepsilon, && |I|+|J|\leq N,\\
\|\partial^I L^J \phi\|_{L^2(\mathcal{H}_2^*)} + \|\partial_t\partial^I L^J \phi\|_{L^2(\mathcal{H}_2^*)} &\leq C_0\,\varepsilon, && |I|+|J|\leq N.
\end{aligned}
$$

We outline here the bootstrap argument and refer to [39, Section 2.4] for further details. Throughout we fix a sufficiently large integer N and we proceed by assuming that the following energy bounds have been established within a hyperbolic time interval $[2, s^*]$:

$$
\begin{aligned}
E_M(s, \partial^I L^J h_{\alpha\beta})^{1/2} &\leq C_1\varepsilon s^\delta, && N-3\leq |I|+|J|\leq N,\\
E_{M,c^2}(s, \partial^I L^J \phi)^{1/2} &\leq C_1\varepsilon s^{1/2+\delta}, && N-3\leq |I|+|J|\leq N,
\end{aligned}
\tag{2.11a}
$$

and

$$
E_M(s, \partial^I L^J h_{\alpha\beta})^{1/2} + E_{M,c^2}(s, \partial^I L^J \phi)^{1/2} \leq C_1\varepsilon s^\delta, \quad |I|+|J|\leq N-4.
\tag{2.11b}
$$

More precisely, we set

$$
s^* := \sup\Big\{s_1 \,\Big|\, \text{for all } 2\leq s\leq s_1, \text{the bounds (2.11) hold}\Big\}.
$$

Since standard arguments for local existence do apply (see [39, Chap. 11]) and, clearly, s^* is not trivial in the sense that, if we choose $C_1 > C_0$, then by continuity we have $s^* > 2$.

By continuity, when $s = s^*$ at least one of the following equalities holds:

$$
\begin{aligned}
E_M(s, \partial^I L^J h_{\alpha\beta})^{1/2} &= C_1\varepsilon s^\delta, && N-3\leq |I|+|J|\leq N,\\
E_{M,c^2}(s, \partial^I L^J \phi)^{1/2} &= C_1\varepsilon s^{1/2+\delta}, && N-3\leq |I|+|J|\leq N,\\
E_M(s, \partial^I L^J h_{\alpha\beta})^{1/2} + E_{M,c^2}(s, \partial^I L^J \phi)^{1/2} &= C_1\varepsilon s^\delta, && |I|+|J|\leq N-4.
\end{aligned}
\tag{2.12}
$$

Our main task for the rest of this work is to derive from (2.11) the *improved energy bounds*:

$$
\begin{aligned}
&E_M(s, \partial^I L^J h_{\alpha\beta})^{1/2} \leq \frac{1}{2} C_1 \varepsilon s^\delta, && N-3 \leq |I| + |J| \leq N, \\
&E_{M,c^2}(s, \partial^I L^J \phi)^{1/2} \leq \frac{1}{2} C_1 \varepsilon s^{1/2+\delta}, && N-3 \leq |I| + |J| \leq N, \\
&E_M(s, \partial^I L^J h_{\alpha\beta})^{1/2} + E_{M,c^2}(s, \partial^I L^J \phi)^{1/2} \leq \frac{1}{2} C_1 \varepsilon s^\delta, && |I| + |J| \leq N-4.
\end{aligned}
\tag{2.13}
$$

By comparing with (2.12), we will be able to conclude that the interval $[2, s^*]$ extends to the maximal time of existence of the local solution. Then by a standard local existence argument, this local solution extends to all time values s.

2.5 Outline of the Monograph

We must therefore derive the improved energy bounds (2.13) and, to this end, the rest of this work is organized as follows. In Chapter 3, we begin by presenting various analytical tools which are required for the analysis of (general functions or) solutions defined on the hyperboloidal foliation. In particular, we establish first an energy estimate for wave equations and Klein-Gordon equations on a curved spacetime, then a sup-norm estimate based on characteristic integration, and next sharp L^∞–L^∞ estimates for wave equations and for Klein-Gordon equations, as well as Sobolev and Hardy inequalities on hyperboloids.

In Chapter 4, we discuss the reduction of the Einstein-massive field system and we establish the quasi-null structure in wave gauge. We provide a classification of all relevant nonlinearities arising in the problem and we carefully study the nonlinear structure of the Einstein equations in the semi-hyperboloidal frame.

Next, in Chapter 5 we formulate our full list of bootstrap assumptions and we write down basic estimates that directly follow from these assumptions. In Chapter 6, we are in a position to provide a preliminary control of the nonlinearities of the Einstein equations in the L^2 and L^∞ norms. In Chapter 7, we establish estimates which are tight to the wave gauge condition.

An estimate of the second-order derivatives of the metric coefficients is then derived in Chapter 8, while in Chapter 9 we obtain a sup-norm estimate based on integration on characteristics and we apply it to the control of quasi-null terms.

We are then able, in Chapter 10, to derive the low-order "refined" energy estimate for the metric and next, in Chapter 11, to control the low-order sup-norm of the metric as well as of the scalar field. In Chapter 12, we improve our bound on the high-order energy for the metric components and the scalar field. In Chapter 13, based on this improved energy bound at high-order, we establish high-order sup-norm estimates. Finally, in Chapter 14, we improve the low-order energy bound on the scalar field and we conclude our bootstrap argument.

Chapter 3

Functional Analysis on Hyperboloids of Minkowski Spacetime

3.1 Energy estimate on hyperboloids

In this section, we need to adapt the techniques we introduced earlier in [39; 41] to the compact Schwarzschild perturbations under consideration in the present Monograph, since these techniques were established for compactly supported initial data. Here, the initial data is not supported in the unit ball but coincides with Schwarzschild space outside the unit ball. As mentioned in the previous section, the curved part of the metric (for a solution of the Einstein-massive field system with a compact Schwarzschild perturbation) is not compactly supported in the light-cone $\mathcal{K}$, while the hyperboloidal energy estimate developed in [39] were assuming this. Therefore, we need to revisit the energy estimate and take suitable boundary terms into account.

Proposition 3.1 (Energy estimate. I). *Let $(h_{\alpha\beta}, \phi)$ be a solution of the Einstein-massive field system associated with an initial data set that is a compact Schwarzschild perturbation with mass $m_S \in (0,1)$. Assume that there exists a constant $\kappa > 1$ such that*

$$\kappa^{-1} E_M^*(s,u)^{1/2} \leq E_g^*(s,u)^{1/2} \leq \kappa E_M^*(s,u)^{1/2}. \tag{3.1}$$

Then, there exists a positive constant C (depending upon N and κ) such that *the following energy estimate holds (for all $\alpha, \beta \leq 3$, and $|I| + |J| \leq N$):*

$$
\begin{aligned}
&E_M(s, \partial^I L^J h_{\alpha\beta})^{1/2} \\
&\leq C E_g(2, \partial^I L^J h_{\alpha\beta})^{1/2} + C m_S + C \int_2^s \|\partial^I L^J F_{\alpha\beta}\|_{L^2(\mathcal{H}_\tau^*)} d\tau \\
&\quad + C \int_2^s \|[\partial^I L^J, H^{\mu\nu}\partial_\mu\partial_\nu] h_{\alpha\beta}\|_{L^2(\mathcal{H}_\tau^*)} d\tau + C \int_2^s M_{\alpha\beta}[\partial^I L^J h](\tau)\, d\tau \\
&\quad + C \int_2^s \Big(\|\partial^I L^J (\partial_\alpha\phi\partial_\beta\phi)\|_{\mathcal{H}_\tau^*} + \|\partial^I L^J (\phi^2 g_{\alpha\beta})\|_{\mathcal{H}_\tau^*} \Big)\, d\tau,
\end{aligned}
\tag{3.2}
$$

in which $M_{\alpha\beta}[\partial^I L^J h](s)$ is a positive function such that

$$
\begin{aligned}
&\int_{\mathcal{H}_s^*} (s/t) \big| \partial_\mu g^{\mu\nu} \partial_\nu (\partial^I L^J h_{\alpha\beta}) \partial_t (\partial^I L^J h_{\alpha\beta}) \\
&\quad - \frac{1}{2} \partial_t g^{\mu\nu} \partial_\mu (\partial^I L^J h_{\alpha\beta}) \partial_\nu (\partial^I L^J h_{\alpha\beta}) \big| \, dx \\
&\leq M_{\alpha\beta}[\partial^I L^J h](s) E_M^*(s, \partial^I L^J h_{\alpha\beta})^{1/2}.
\end{aligned}
\tag{3.3}
$$

The proof of this estimate is done as follows: in the exterior part of the hyperboloid (i.e. $\mathcal{H}_s \cap \mathcal{K}^c$), the metric coincides with the Schwarzschild metric and we can calculate the energy by an explicit expression. On the other hand, the interior part is bounded as follows.

Lemma 3.1. *Under the assumptions in Proposition 3.1, one has*

$$
\begin{aligned}
&E_M^*(s, \partial^I L^J h_{\alpha\beta})^{1/2} \\
&\leq C E_g^*(2, \partial^I L^J h_{\alpha\beta})^{1/2} + C m_S + C \int_2^s M_{\alpha\beta}(\tau, \partial^I L^J h_{\alpha\beta})\, d\tau \\
&\quad + C \int_2^s \|\partial^I L^J F_{\alpha\beta}\|_{L^2(\mathcal{H}_\tau^*)} d\tau + C \int_2^s \|[\partial^I L^J, H^{\mu\nu}\partial_\mu\partial_\nu] h_{\alpha\beta}\|_{L^2(\mathcal{H}_\tau^*)} d\tau \\
&\quad + C \int_2^s \big(\|\partial^I L^J (\partial_\alpha\phi\partial_\beta\phi)\|_{L^2(\mathcal{H}_\tau^*)} + \|\partial^I L^J (\phi^2 g_{\alpha\beta})\|_{L^2(\mathcal{H}_\tau^*)} \big) d\tau.
\end{aligned}
\tag{3.4}
$$

Proof. We consider the wave equation $g^{\mu\nu}\partial_\mu\partial_\nu h_{\alpha\beta} = F_{\alpha\beta} - 16\pi\partial_\alpha\phi\partial_\beta\phi - 8\pi c^2\phi^2 g_{\alpha\beta}$ satisfied by the curved part of the metric and differentiate it (with $\partial^I L^J$ with $|I| + |J| \leq N$):

$$
\begin{aligned}
g^{\mu\nu}\partial_\mu\partial_\nu \partial^I L^J h_{\alpha\beta} = &- [\partial^I L^J, H^{\mu\nu}\partial_\mu\partial_\nu] h_{\alpha\beta} + \partial^I L^J F_{\alpha\beta} \\
&- 16\pi \partial^I L^J (\partial_\alpha\phi\partial_\beta\phi) - 8\pi c^2 \partial^I L^J (\phi^2 g_{\alpha\beta}).
\end{aligned}
$$

Using the multiplier $-\partial_t\partial^I L^J h_{\alpha\beta}$, we obtain the general identity

$$
\begin{aligned}
&\partial_t\big(- (1/2)g^{00}|\partial_t\partial^I L^J h_{\alpha\beta}|^2 + (1/2)g^{ab}\partial_a\partial^I L^J h_{\alpha\beta}\partial_b\partial^I L^J h_{\alpha\beta}\big)\\
&\quad - \partial_a\big(g^{a\nu}\partial_\nu\partial^I L^J h_{\alpha\beta}\partial_t\partial^I L^J h_{\alpha\beta}\big)\\
&= \frac{1}{2}\partial_t g^{\mu\nu}\partial_\mu\partial^I L^J h_{\alpha\beta} - \partial_\mu g^{\mu\nu}\partial_t\partial^I L^J h_{\alpha\beta}\partial_\nu\partial^I L^J h_{\alpha\beta}\\
&\quad + [\partial^I L^J, H^{\mu\nu}\partial_\mu\partial_\nu]h_{\alpha\beta}\partial_t\partial^I L^J h_{\alpha\beta} - \partial^I L^J F_{\alpha\beta}\partial_t\partial^I L^J h_{\alpha\beta}\\
&\quad + 16\pi\partial^I L^J\big(\partial_\alpha\phi\partial_\beta\phi\big)\partial_t\partial^I L^J h_{\alpha\beta} + 8\pi c^2\partial^I L^J\big(\phi^2 g_{\alpha\beta}\big)\partial_t\partial^I L^J h_{\alpha\beta}.
\end{aligned}
\tag{3.5}
$$

For simplicity, we write $u = \partial^I L^J h_{\alpha\beta}$ and $W := \big(- (1/2)g^{00}|\partial_t u|^2 + (1/2)g^{ab}\partial_a u\partial_b u, -g^{a\nu}\partial_\nu u\partial_t u\big)$ for the energy flux, while

$$
\begin{aligned}
\mathcal{F} :=& \frac{1}{2}\partial_t g^{\mu\nu}\partial_\mu\partial^I L^J h_{\alpha\beta} - \partial_\mu g^{\mu\nu}\partial_t\partial^I L^J h_{\alpha\beta}\partial_\nu\partial^I L^J h_{\alpha\beta}\\
&+ [\partial^I L^J, H^{\mu\nu}\partial_\mu\partial_\nu]h_{\alpha\beta}\partial_t\partial^I L^J h_{\alpha\beta} - \partial^I L^J F_{\alpha\beta}\partial_t\partial^I L^J h_{\alpha\beta}\\
&+ 16\pi\partial^I L^J\big(\partial_\alpha\phi\partial_\beta\phi\big)\partial_t\partial^I L^J h_{\alpha\beta} + 8\pi c^2\partial^I L^J\big(\phi^2 g_{\alpha\beta}\big)\partial_t\partial^I L^J h_{\alpha\beta}.
\end{aligned}
$$

Then, by defining Div with respect to the Euclidian metric on $\mathbb{R}^{3+1}$, (3.5) reads $\mathrm{Div}W = \mathcal{F}$ and we can next integrate this equation in the region $\mathcal{K}_{[2,s]}$ and write $\int_{\mathcal{K}_{[2,s]}} \mathrm{Div}W dxdt = \int_{\mathcal{K}[2,s]}\mathcal{F}dxdt$. In the left-hand side, we apply Stokes' formula:

$$
\int_{\mathcal{K}_{[2,s]}} \mathrm{Div}W dxdt = \int_{\mathcal{H}^*_s} W\cdot n d\sigma + \int_{\mathcal{H}^*_2} W\cdot n d\sigma + \int_{B_{[2,s]}} W\cdot n d\sigma,
$$

where $B_{[2,s]}$ is the boundary of $\mathcal{K}_{[2,s]}$, which is $\big\{(t,x)|t = r+1, 3/2\leq r\leq (s^2-1)/2\big\}$. An easy calculation shows that

$$
\begin{aligned}
&\int_{\mathcal{K}_{[2,s]}} \mathrm{Div}W dxdt\\
&= \frac{1}{2}\Big(E^*_g(s,\partial^I L^J h_{\alpha\beta}) - E^*_g(2,\partial^I L^J h_{\alpha\beta})\Big)\\
&\quad + \int_{3/2\leq r\leq(s^2-1)/2}\int_{\mathbb{S}^2} W\cdot(-\sqrt{2}/2, \sqrt{2}x^a/2r)\sqrt{2}r^2 drd\omega ds,
\end{aligned}
\tag{3.6}
$$

where $d\omega$ is the standard Lebesgue measure on $\mathbb{S}^2$. Recall that $g_{\alpha\beta} = g_{S\alpha\beta}$ in a neighborhood of $B_{[2,s]}$. An explicit calculation shows that $W = \big((1/2){g_S}^{ab}\partial_a\partial^I L^J h_{S\alpha\beta}\partial_b\partial^I L^J h_{S\alpha\beta}, 0\big)$ on $B_{[2,s]}$. We have

$$
\begin{aligned}
&\int_{3/2\leq r\leq(s^2-1)/2}\int_{\mathbb{S}^2} W\cdot(-\sqrt{2}/2, \sqrt{2}x^a/2r)\sqrt{2}r^2 drd\omega\\
&= -2\pi\int_{3/2}^{(s^2-1)/2} {g_S}^{ab}\partial_a\partial^I L^J h_{S\alpha\beta}\partial_b\partial^I L^J h_{S\alpha\beta}r^2 drds
\end{aligned}
$$

with $h_{S\alpha\beta} := g_{S\alpha\beta} - m_{\alpha\beta}$. This leads us to

$$\frac{d}{ds}\int_{B_{[2,s]}} W\cdot n d\sigma = -\frac{\pi}{2}s(s^2-1)^2 {g_S}^{ab}\partial_a\partial^I L^J h_{S\alpha\beta}\partial_b\partial^I L^J h_{S\alpha\beta}\Big|_{r=\frac{s^2-1}{2}}.$$

Assuming that m_S is sufficiently small, we see that

$$\left|{g_S}^{ab}\partial_a\partial^I L^J h_{S\alpha\beta}\partial_b\partial^I L^J h_{S\alpha\beta}\right| \leq C m_S^2 r^{-4} \leq C m_S^2 s^{-8}, \qquad 3/2 \leq r.$$

We have

$$\left|\frac{d}{ds}\int_{B_{[2,s]}} W\cdot n d\sigma\right| \leq C m_S^2 s^{-3}. \tag{3.7}$$

Now, we combine $\mathrm{Div}W = \mathcal{F}$ and (3.6) and differentiate in s:

$$\frac{1}{2}\frac{d}{ds}E_g^*(s,\partial^I L^J h_{\alpha\beta}) + \frac{d}{ds}\int_{B_{[2,s]}} W\cdot n d\sigma = \frac{d}{ds}\int_{\mathcal{K}_{[2,s]}} \mathcal{F}\, dxdt,$$

which leads us to

$$\begin{aligned}
&E_g^*(s,\partial^I L^J h_{\alpha\beta})^{1/2}\frac{d}{ds}\left(E_g^*(s,\partial^I L^J h_{\alpha\beta})^{1/2}\right)\\
&= -\frac{d}{ds}\int_{B_{[2,s]}} W\cdot n d\sigma + \frac{d}{ds}\int_2^s\int_{\mathcal{H}_s^*}(s/t)\mathcal{F}\, dxds.
\end{aligned}$$

Then, in view of (3.7) we have

$$E_g^*(s,\partial^I L^J h_{\alpha\beta})^{1/2}\frac{d}{ds}\left(E_g^*(s,\partial^I L^J h_{\alpha\beta})^{1/2}\right) \leq \int_{\mathcal{H}_s^*}(s/t)|\mathcal{F}|\, dx + C m_S^2 s^{-3}. \tag{3.8}$$

In view of the notation and assumptions in Proposition 3.1, we have

$$\begin{aligned}
&\int_{\mathcal{H}_s^*}|(s/t)\mathcal{F}|\, dx \leq \int_{\mathcal{H}_s^*}|(s/t)\partial_t\partial^I L^J h_{\alpha\beta}\partial^I L^J F_{\alpha\beta}|dx\\
&+\int_{\mathcal{H}_s^*}|(s/t)\partial_t\partial^I L^J h_{\alpha\beta}[\partial^I L^J, H^{\mu\nu}\partial_\mu\partial_\nu]h_{\alpha\beta}|dx\\
&+16\pi\int_{\mathcal{H}_s^*}|(s/t)\partial_t\partial^I L^J h_{\alpha\beta}\partial^I L^J(\partial_\alpha\phi\partial_\beta\phi)|\, dx\\
&+8\pi c^2\int_{\mathcal{H}_s^*}|(s/t)\partial_t\partial^I L^J h_{\alpha\beta}\partial^I L^J(\phi^2 g_{\alpha\beta})|\, dx + M[\partial^I L^J h](s)E_M^*(s,\partial^I L^J h_{\alpha\beta})^{1/2}\\
&\leq \|(s/t)\partial_t\partial^I L^J h_{\alpha\beta}\|_{L^2(\mathcal{H}_s^*)}\left(\|\partial^I L^J F_{\alpha\beta}\|_{L^2(\mathcal{H}_s^*)} + \|\partial^I L^J, [H^{\mu\nu}\partial_\mu\partial_\nu]h_{\alpha\beta}\|_{L^2(\mathcal{H}_s^*)}\right)\\
&\quad + C\|(s/t)\partial_t\partial^I L^J h_{\alpha\beta}\|_{L^2(\mathcal{H}_s^*)}\left(\|\partial^I L^J(\partial_\alpha\phi\partial_\beta\phi)\|_{L^2(\mathcal{H}_s^*)} + \|\partial^I L^J(\phi^2 g_{\alpha\beta})\|_{L^2(\mathcal{H}_s^*)}\right)\\
&\quad + M[\partial^I L^J h](s)E_M^*(s,\partial^I L^J h_{\alpha\beta})^{1/2},
\end{aligned}$$

so that

$$\int_{\mathcal{H}_s^*} |(s/t)\mathcal{F}|\,dx$$
$$\le CE_M^*(s,\partial^I L^J h_{\alpha\beta})^{1/2}\Big(\|\partial^I L^J F_{\alpha\beta}\|_{L^2(\mathcal{H}_s^*)} + \|\partial^I L^J, [H^{\mu\nu}\partial_\mu\partial_\nu]h_{\alpha\beta}\|_{L^2(\mathcal{H}_s^*)}$$
$$+ \|\partial^I L^J(\partial_\alpha\phi\partial_\beta\phi)\|_{L^2(\mathcal{H}_s^*)} + \|\partial^I L^J(\phi^2 g_{\alpha\beta})\|_{L^2(\mathcal{H}_s^*)} + M[\partial^I L^J h](s)\Big).$$

For simplicity, we write

$$L(s) := \|\partial^I L^J F_{\alpha\beta}\|_{L^2(\mathcal{H}_s^*)} + \|\partial^I L^J, [H^{\mu\nu}\partial_\mu\partial_\nu]h_{\alpha\beta}\|_{L^2(\mathcal{H}_s^*)}$$
$$+ \|\partial^I L^J(\partial_\alpha\phi\partial_\beta\phi)\|_{L^2(\mathcal{H}_s^*)} + \|\partial^I L^J(\phi^2 g_{\alpha\beta})\|_{L^2(\mathcal{H}_s^*)} + M[\partial^I L^J h](s)$$

and $y(s) := E_g^*(s,\partial^I L^J h_{\alpha\beta})^{1/2}$. In view of (3.1), we have

$$E_M^*(s,\partial^I L^J h_{\alpha\beta})^{1/2} \le C\kappa E_g^*(s,\partial^I L^J h_{\alpha\beta})^{1/2}$$

and (3.8) leads us to $y(s)y'(s) = C\kappa y(s)L(s) + Cm_S^2 s^{-3}$. By Lemma 3.2 stated shortly below, we conclude that (with $m_S = \varepsilon$ and $\sigma = 2$ therein)

$$y(s) \le y(0) + Cm_S + C\kappa \int_2^s L(s)ds.$$

By recalling (3.1), the above inequality leads us to (3.4). □

Lemma 3.2. *The nonlinear inequality $y(\tau)y'(\tau) \le g(\tau)y(\tau) + C^2\varepsilon^2\tau^{-1-\sigma}$, in which the function $y : [2,s] \to \mathbb{R}^+$ is sufficiently regular, the function g is positive and locally integrable, and C,ε,σ are positive constants, implies the linear inequality*

$$y(\tau) \le y(2) + C\varepsilon\left(1+\sigma^{-1}\right) + \int_2^\tau g(\eta)d\eta.$$

Proof. We denote by $I = \{\tau \in [2,s] | y(s) > C\varepsilon\}$. In view of the continuity of y, $I = \bigcup_{i\in\mathbb{N}}(I_n \cap [2,s])$ where I_n are open intervals disjoint from each other. For $\tau \notin I$, $y(\tau) \le C\varepsilon$. For $\tau \in I$, there exists some integer i such that $\tau \in I_i \cap [2,s]$. Let $\inf(I_i \cap [2,s]) = s_0 \ge 2$, then on $I_n \cap [2,s]$,

$$y'(\tau) \le g(\tau) + \frac{C^2\varepsilon^2\tau^{-1-\sigma}}{y(\tau)} \le g(\tau) + C\varepsilon\tau^{-1-\sigma}.$$

This leads us to

$$\int_{s_0}^\tau y'(\eta)d\eta \le \int_{s_0}^\tau g(\eta)d\eta + C\varepsilon\int_{s_0}^\tau s^{-1-\sigma}ds$$
$$\le \int_2^\tau g(\eta)d\eta + C\varepsilon\int_2^\infty s^{-1-\sigma}ds \le \int_2^\tau g(\eta)d\eta + C\varepsilon\sigma^{-1}$$

and $y(\tau) - y(s_0) \leq \int_2^\tau g(\eta)d\eta + C\varepsilon\sigma^{-1}$. By continuity, either $s_0 \in (2, s)$ which leads us to $y(s_0) = C\varepsilon$, or else $s_0 = 2$ which leads us to $y(s_0) = y(2)$. Then, we obtain

$$y(\tau) \leq \max\{y(2), C\varepsilon\} + C\varepsilon\sigma^{-1} + \int_2^\tau g(\eta)d\eta.$$

□

To complete the proof of Proposition 3.1, we need the following additional observation, which is checked by an explicit calculation (omitted here).

Lemma 3.3. *The following uniform estimate holds (for all a, α, β, all relevant I, J, and for some $C = C(I, J)$)*

$$\int_{\mathcal{H}_s \cap \mathcal{K}^c} |\underline{\partial}_a \partial^I L^J h_{S\alpha\beta}|^2 dx + \int_{\mathcal{H}_s \cap \mathcal{K}^c} (s/t)|\partial_t \partial^I L^J h_{S\alpha\beta}|^2 dx \leq C m_S^2. \quad (3.9)$$

Proof of Proposition 3.1. We observe that

$$\begin{aligned} &E_g(s, \partial^I L^J h_{\alpha\beta}) \\ &\leq E_g^*(s, \partial^I L^J h_{\alpha\beta}) + C \int_{\mathcal{H}_s \cap \mathcal{K}^c} |\underline{\partial}_a \partial^I L^J h_{S\alpha\beta}|^2 dx \\ &\qquad + \int_{\mathcal{H}_s \cap \mathcal{K}^c} (s/t)|\partial_t \partial^I L^J h_{S\alpha\beta}|^2 dx. \end{aligned}$$

Combining (3.4) with Lemma 3.3 allows us to complete the proof of (3.2). □

For all solutions to the Einstein-massive field system associated with compact Schwarzschild perturbations, the scalar field ϕ is also supported in $\mathcal{K}$. So the energy estimate for ϕ remains identical to the one in [41].

Proposition 3.2 (Energy estimate. II). *Under the assumptions in Proposition 3.1, the scalar field ϕ satisfies*

$$\begin{aligned} E_{M,c^2}(s, \partial^I L^J \phi)^{1/2} &\leq C E_{g,c^2}(2, \partial^I L^J \phi)^{1/2} \\ &\quad + \int_2^s \big|[\partial^I L^J, H^{\mu\nu}\partial_\mu\partial_\nu]\phi\big| d\tau + \int_2^s M[\partial^I L^J \phi](\tau)\, d\tau, \end{aligned} \quad (3.10)$$

in which $M[\partial^I L^J \phi](s)$ denotes a positive function such that

$$\begin{aligned} &\int_{\mathcal{H}_s} (s/t)\left|\partial_\mu g^{\mu\nu} \partial_\nu(\partial^I L^J \phi)\partial_t(\partial^I L^J \phi) - \frac{1}{2}\partial_t g^{\mu\nu}\partial_\mu(\partial^I L^J \phi)\partial_\nu(\partial^I L^J \phi)\right| dx \\ &\leq M[\partial^I L^J \phi](s) E_{M,c^2}(s, \partial^I L^J \phi)^{1/2}. \end{aligned} \quad (3.11)$$

3.2 Sup-norm estimate based on curved characteristic integration

We now revisit an important technical tool introduced first in Lindblad and Rodnianski [45]. This is an L^∞ estimate on the gradient of solutions to a wave equation posed in a curved background. For our problem, we must adapt this tool to the hyperboloidal foliation and we begin by stating without proof the following identity.

Lemma 3.4. (Decomposition of the flat wave operator in the null frame). *For every smooth function u, the following identity holds:*

$$-\Box u = r^{-1}(\partial_t + \partial_r)(\partial_t - \partial_r)(ru) - \sum_{a<b} \big(r^{-1}\Omega_{ab}\big)^2 u \tag{3.12}$$

with $\Omega_{ab} = x^a\partial_b - x^b\partial_a = x^a\underline{\partial}_b - x^b\underline{\partial}_a$ (defined earlier).

We then write $\partial_t = \frac{t}{t+r}(\partial_t - \partial_r) + \frac{x^a t}{(t+r)r}\underline{\partial}_a$ and thus

$$\begin{aligned}\partial_t\partial_t &= \frac{t^2}{(t+r)^2}(\partial_t - \partial_r)^2 + \frac{t}{t+r}(\partial_t - \partial_r)\left(\frac{x^a t\underline{\partial}_a}{r(t+r)}\right)\\ &\quad + \frac{x^a t}{r(t+r)}\underline{\partial}_a\left(\frac{t}{t+r}(\partial_t - \partial_r)\right) + \left(\frac{x^a t}{r(t+r)}\underline{\partial}_a\right)^2 + \frac{\partial_t - \partial_r}{t+r}.\end{aligned}$$

Consequently, we have found the decomposition

$$\begin{aligned}r\partial_t\partial_t u &= \frac{t^2}{(t+r)^2}(\partial_t - \partial_r)^2(ru) + \frac{2t^2}{(t+r)^2}(\partial_t - \partial_r)u\\ &\quad + \frac{rt}{t+r}(\partial_t - \partial_r)\left(\frac{x^a t}{r(t+r)}\underline{\partial}_a u\right)\\ &\quad + \frac{x^a t}{(t+r)}\underline{\partial}_a\left(\frac{t}{t+r}(\partial_t - \partial_r)u\right) + r\left(\frac{x^a t}{r(t+r)}\underline{\partial}_a\right)^2 u + \frac{r(\partial_t - \partial_r)u}{t+r}\\ &=: \frac{t^2}{(t+r)^2}(\partial_t - \partial_r)^2(ru) + W_1[u].\end{aligned} \tag{3.13}$$

On the other hand, the curved part of the reduced wave operator $H^{\alpha\beta}\partial_\alpha\partial_\beta$ can be decomposed in the semi-hyperboloidal frame as follows:

$$\begin{aligned}H^{\alpha\beta}\partial_\alpha\partial_\beta u &= \underline{H}^{\alpha\beta}\underline{\partial}_\alpha\underline{\partial}_\beta u + H^{\alpha\beta}\partial_\alpha\big(\Psi_\beta^{\beta'}\big)\underline{\partial}_{\beta'} u\\ &= \underline{H}^{00}\partial_t\partial_t u + \underline{H}^{a0}\underline{\partial}_a\partial_t u + \underline{H}^{0a}\partial_t\underline{\partial}_a u + \underline{H}^{ab}\underline{\partial}_a\underline{\partial}_b u + H^{\alpha\beta}\partial_\alpha\big(\Psi_\beta^{\beta'}\big)\underline{\partial}_{\beta'} u.\end{aligned}$$

The "good" part of the curved wave operator (i.e. terms containing one derivative tangential to the hyperboloids) is defined to be

$$R[u,H] := \underline{H}^{a0}\underline{\partial}_a\partial_t u + \underline{H}^{0a}\partial_t\underline{\partial}_a u + \underline{H}^{ab}\underline{\partial}_a\underline{\partial}_b u + H^{\alpha\beta}\partial_\alpha\big(\Psi_\beta^{\beta'}\big)\underline{\partial}_{\beta'} u, \tag{3.14}$$

and, with this notation together with (3.13),

$$r H^{\alpha\beta}\partial_\alpha\partial_\beta u = \frac{t^2 \underline{H}^{00}}{(t+r)^2}(\partial_t - \partial_r)\big((\partial_t - \partial_r)(ru)\big) + \underline{H}^{00} W_1[u] + rR[u,H]. \tag{3.15}$$

Then, by combining (3.12) for the flat wave operator and (3.15) for the curved part, we reach the following conclusion.

Lemma 3.5 (Decomposition of the reduced wave operator $\widetilde{\Box}_g$). *Let u be a smooth function defined in $\mathbb{R}^{3+1}$ and $H^{\alpha\beta}$ be functions in $\mathbb{R}^{3+1}$. Then the following identity holds:*

$$\begin{aligned}&\Big((\partial_t + \partial_r) - t^2(t+r)^{-2}\underline{H}^{00}(\partial_t - \partial_r)\Big)\Big((\partial_t - \partial_r)(ru)\Big)\\ &= -r\,\widetilde{\Box}_g u + r\sum_{a<b}\left(r^{-1}\Omega_{ab}\right)^2 u + \underline{H}^{00} W_1[u] + rR[u,H]\end{aligned} \tag{3.16}$$

with the notation above.

Now we are ready to establish the desired estimate of this section. For convenience, we set

$$\mathcal{K}^{\rm int} := \big\{(t,x) | r \leq \frac{3}{5} t\big\} \cap \mathcal{K}, \quad \mathcal{K}^{\rm int}_{[s_0,s_1]} := \big\{(t,x) \in \mathcal{K}^{\rm int} \,/\, s_0^2 \leq t^2 - r^2 \leq s_1^2\big\}$$

and we denote by $\partial_B \mathcal{K}^{\rm int}_{[s_0,s_1]}$ the following "boundary" of $\mathcal{K}^{\rm int}_{[s_0,s_1]}$

$$\partial_B \mathcal{K}^{\rm int}_{[s_0,s_1]} := \big\{(t,x) \,/\, r = (3/5)t,\ (5/4)s_0 \leq t \leq (5/4)s_1\big\}.$$

We will now prove the following sharp decay property for solutions to the wave equation on a curved spacetime.

Proposition 3.3. (Sup-norm estimate based on characteristic integration). *Let u be a solution to the wave equation on curved spacetime $-\Box u - H^{\alpha\beta}\partial_\alpha\partial_\beta u = F$, where $H^{\alpha\beta}$ are given functions. Given any point (t_0, x_0), denote by $(t, \varphi(t; t_0, x_0))$ the integral curve of the vector field*

$$\partial_t + \frac{(t+r)^2 + t^2\underline{H}^{00}}{(t+r)^2 - t^2\underline{H}^{00}}\partial_r$$

passing through (t_0, x_0), that is, $\varphi(t_0; t_0, x_0) = x_0$. Then, there exist two positive constants ε_s and $a_0 \geq 2$ such that for $t \geq a_0$

$$|\underline{H}^{00}| \leq \varepsilon_s (t-r)/t, \tag{3.17}$$

then for all $s \geq a_0$ and $(t,x) \in \mathcal{K}\backslash\mathcal{K}^{int}_{[2,s]}$ one has

$$|(\partial_t - \partial_r)u(t,x)| \leq t^{-1} \sup_{\partial_B \mathcal{K}^{int}_{[2,s]} \cup \partial\mathcal{K}} \Big(|(\partial_t - \partial_r)(ru)|\Big) + Ct^{-1}|u(t,x)|$$
$$+ t^{-1} \int_{a_0}^{t} \tau |F(\tau, \varphi(\tau;t,x))| d\tau + t^{-1} \int_{a_0}^{t} |M_s[u,H]|_{(\tau,\varphi(\tau;t,x))} d\tau, \tag{3.18}$$

where $F = -\Box u - H^{\alpha\beta}\partial_\alpha\partial_\beta u$ is the right-hand side of the wave equation,

$$M_s[u,H] := r \sum_{a<b} \left(r^{-1}\Omega_{ab}\right)^2 u + \underline{H}^{00} W_1[u] + rR[u,H],$$

in which one can guarantee that the associated integral curve satisfies $(\tau, \varphi(\tau;t,x)) \in \mathcal{K}\backslash\mathcal{K}^{int}_{[2,s]}$ for $2 \leq a_0 < \tau < t$, but $(a_0, \varphi(a_0;t,x)) \in \partial_B\mathcal{K}^{int}_{[2,s_0]} \cup \partial\mathcal{K}$ at the initial time a_0.

Proof. Under the condition (3.17), the decomposition (3.16) can be rewritten in the form

$$\begin{aligned} &\left(\partial_t + \frac{1+t^2(t+r)^{-2}\underline{H}^{00}}{1-t^2(t+r)^{-2}\underline{H}^{00}}\partial_r\right)\big((\partial_t - \partial_r)(ru)\big) =: \mathcal{L}\big((\partial_t - \partial_r)(ru)\big) \\ &= \frac{-r\widetilde{\Box}_g u + r\sum_{a<b}\left(r^{-1}\Omega_{ab}\right)^2 u + \underline{H}^{00}W_1[u] + rR[u,H]}{1-t^2(t+r)^{-2}\underline{H}^{00}} =: \mathcal{F}. \end{aligned} \tag{3.19}$$

In other words, (3.19) reads $\mathcal{L}\big((\partial_r - \partial_r)(ru)\big) = \mathcal{F}$ and by writing

$$v_{t_0,x_0}(t) := \big((\partial_r - \partial_r)(ru)\big)(t, \varphi(t;t_0,x_0)),$$

we have

$$\frac{d}{dt}v_{t_0,x_0}(t) = \mathcal{L}\big((\partial_t - \partial_r)(ru)\big)(t,\varphi(t;t_0,x_0)) = \mathcal{F}(t,\varphi(t;t_0,x_0)).$$

By integration, we have $v_{t_0,x_0}(t_0) = v_{t_0,x_0}(a) + \int_a^{t_0} \mathcal{F}(t,\varphi(t;t_0,x_0))\, dt$.

Fix $s_0^2 = t_0^2 - r_0^2$ with $s_0 > 0$ and take $(t_0,x_0) \in \mathcal{K}_{[2,s]}\backslash\mathcal{K}^{\text{int}}$, that is $\{(t_0,x_0) | (3/5)t_0 \leq r_0 < t_0 - 1\}$. We will prove that there exists some $a \geq 2$ such that for all $t \in [a,t_0]$, $(t,\varphi(t;t_0,x_0)) \in \mathcal{K}_{[2,s]}\backslash\mathcal{K}^{\text{int}}$ and $(a,\varphi(a;t_0,x_0)) \in \partial_B\mathcal{K}^{\text{int}}_{[2,s_0]} \cup \partial\mathcal{K}$, that is, for $t < t_0$, $(t,\varphi(t;t_0,x_0))$ will not intersect $\mathcal{H}_{s_0}$ again before leaving the region $\mathcal{K}_{[2,s_0]}\backslash\mathcal{K}^{\text{int}}$. This is due to the following observation: denote by $|\varphi(t;t_0,x_0)|$ the Euclidian norm of $\varphi(t;t_0,x_0)$, and by the definition of $\mathcal{L}$, we have

$$\frac{d|\varphi(t;t_0,x_0)|}{dt} = \frac{1+t^2(t+r)^{-2}\underline{H}^{00}}{1-t^2(t+r)^{-2}\underline{H}^{00}}.$$

Also, we observe that for a point (t, x) on the hyperboloid $\mathcal{H}_{s_0}$, we have $r(t) = |x(t)| = \sqrt{t^2 - s_0^2}$, and this leads us to $\frac{dr}{dt} = \frac{t}{r}$. Then we have

$$\begin{aligned}\frac{d\big(|\varphi(t;t_0,x_0)| - r\big)}{dt} &= \frac{1 + t^2(t+r)^{-2}\underline{H}^{00}}{1 - t^2(t+r)^{-2}\underline{H}^{00}} - \frac{t}{r}\\ &= \frac{2t^2(t+r)^{-2}\underline{H}^{00}}{1 - t^2(t+r)^{-2}\underline{H}^{00}} - \frac{t-r}{r}.\end{aligned}$$

So, there exists a constant ε_s such that if $|\underline{H}^{00}| \leq \frac{\varepsilon_s(t-r)}{t}$, then $\frac{d\left(|\varphi(t;t_0,x_0)|-r\right)}{dt} < 0$. Recall that at $t = t_0$, $|\varphi(t_0;t_0,x_0)| = |x_0| = r(t_0)$. We conclude that for all $t < t_0$, $|\varphi(t;t_0,x_0)| > r(t)$ which shows that $(t, \varphi(t;t_0,x_0))$ will never intersect $\mathcal{H}_{s_0}$ again. Furthermore we see that there exists a time a_0 sufficiently small (but still $a_0 \geq 3$) such that $(t, \varphi(t;t_0,x_0))$ leaves $\mathcal{K}_{[2,s]} \backslash \mathcal{K}^{\text{int}}$ by intersecting the boundary $\partial_B \mathcal{K}^{\text{int}}_{[2,s_0]} \cup \partial\mathcal{K}$ at $t = a_0$. So we see that $v_{t_0,x_0}(t_0) = v_{t_0,x_0}(a_0) + \int_{a_0}^{t_0} \mathcal{F}(t, \varphi(t;t_0,x_0))\,dt$, which leads us to

$$\begin{aligned}|v_{t_0,x_0}(t_0)| \leq &\sup_{(t,x)\in\partial_B\mathcal{K}^{\text{int}}_{[2,s_0]}\cup\partial\mathcal{K}} \{|(\partial_t - \partial_r)(ru)|_{(t,x)}|\}\\ &+ \int_2^{t_0} \Big| - r\widetilde{\Box}_g u + r\sum_{a<b}\left(r\Omega_{ab}\right)^2 u + \underline{H}^{00}W_1[u] + rR[u,H]\Big|_{(t,\varphi(t;t_0,x_0))}\,dt.\end{aligned}$$

□

3.3 Sup-norm estimate for wave equations with source

Our sup-norm estimate for the wave equation, established earlier in [41] and based on an explicit formula for solutions (cf. also the Appendix at the end of this monograph), is now revisited and adapted to the problem of compact Schwarzschild perturbations. By applying $\partial^I L^J$ to the Einstein equations (1.6a), we obtain

$$\begin{aligned}\Box\partial^I L^J h_{\alpha\beta} &= -\partial^I L^J\big(H^{\mu\nu}\partial_\mu\partial_\nu h_{\alpha\beta}\big) + \partial^I L^J F_{\alpha\beta}\\ &\quad - 16\pi\partial^I L^J\big(\partial_\alpha\phi\partial_\beta\phi\big) - 8\pi c^2\partial^I L^J\big(\phi^2 g_{\alpha\beta}\big)\\ &=: S^{I,J}_{\alpha\beta} = S^{W,I,J}_{\alpha\beta} + S^{KG,I,J}_{\alpha\beta},\end{aligned} \tag{3.20}$$

with

$$\begin{aligned}S^{W,I,J}_{\alpha\beta} &:= -\partial^I L^J\big(H^{\mu\nu}\partial_\mu\partial_\nu h_{\alpha\beta}\big) + \partial^I L^J F_{\alpha\beta},\\ S^{KG,I,J}_{\alpha\beta} &:= -16\pi\partial^I L^J\big(\partial_\alpha\phi\partial_\beta\phi\big) - 8\pi c^2\partial^I L^J\big(\phi^2 g_{\alpha\beta}\big).\end{aligned}$$

We denote by $\mathbb{1}_\mathcal{K} : \mathbb{R}^4 \to \{0,1\}$ the characteristic function of the set $\mathcal{K}$, and introduce the corresponding decomposition into interior/exterior contributions of the wave source of the Einstein equations:

$$S^{W,I,J}_{\text{Int},\alpha\beta} := \mathbb{1}_\mathcal{K} S^{W,I,J}_{\alpha\beta}, \qquad S^{W,I,J}_{\text{Ext},\alpha\beta} := (1 - \mathbb{1}_\mathcal{K}) S^{W,I,J}_{\alpha\beta},$$

while $S^{KG,I,J}_{\alpha\beta}$ is compactly supported in $\mathcal{K}$ and need not be decomposed. We thus have

$$S^{I,J}_{\alpha\beta} = S^{W,I,J}_{\text{Ext},\alpha\beta} + S^{KG,I,J}_{\alpha\beta} + S^{W,I,J}_{\text{Int},\alpha\beta}. \tag{3.21}$$

Outside the region $\mathcal{K}$, the metric $g_{\alpha\beta}$ coincides with the Schwarzschild metric so that an easy calculation leads us to the following estimate.

Lemma 3.6. *One has* $|S^{W,I,J}_{Ext,\alpha\beta}| \leq Cm_S^2 (1 - \mathbb{1}_\mathcal{K}) r^{-4}$.

We next decompose the initial data for the equations (3.20). Recall that on the initial hypersurface $\{t = 2\}$ and outside the unit ball, the metric coincides with the Schwarzschild metric. We write

$$\partial^I L^J h_{\alpha\beta}(2, \cdot) := I^{0,I,J}_{\text{Int},\alpha} + I^{0,I,J}_{\text{Ext},\alpha\beta},$$
$$I^{0,I,J}_{\text{Int},\alpha} := \widetilde{\chi}(r) \partial^I L^J h_{\alpha\beta}(2, \cdot), \qquad I^{0,I,J}_{\text{Ext},\alpha\beta} := (1 - \widetilde{\chi}(r)) \partial^I L^J h_{\alpha\beta}(2, \cdot),$$

in which $\widetilde{\chi}(\cdot) : \mathbb{R}^+ \to \mathbb{R}^+$ is a smooth cut-off function with

$$\widetilde{\chi}(r) = \begin{cases} 1, & r \leq 1, \\ 0, & r \geq 2. \end{cases}$$

On the other hand, the initial data $\partial_t \partial^I L^J h_{\alpha\beta}(2, \cdot) =: I^1[\partial^I L^J]$ is supported in $\{r \leq 1\}$ since the metric is initially static outside the unit ball. We are in a position to state our main sup-norm estimate.

Proposition 3.4 (Sup-norm estimate for the Einstein equations). *Let $(g_{\alpha\beta}, \phi)$ be a solution of the Einstein-massive field system associated with a compact Schwarzschild initial data. Assume that the source terms in* (3.20) *satisfy*

$$|S^{W,I,J}_{Int,\alpha\beta}| + |S^{KG,I,J}_{\alpha\beta}| \leq C_* t^{-2-\nu} (t - r)^{-1+\mu}. \tag{3.22}$$

Then, when $0 < \mu \leq 1/2$ and $0 < \nu \leq 1/2$, one has

$$|\partial^I L^J h_{\alpha\beta}(t, x)| \leq \frac{CC_*(\alpha, \beta)}{\mu|\nu|} t^{-1} (t - r)^{\mu-\nu} + Cm_S t^{-1}, \tag{3.23}$$

while, when $0 < \mu \leq 1/2$ and $-1/2 \leq \nu < 0$,

$$|\partial^I L^J h_{\alpha\beta}(t, x)| \leq \frac{CC_*(\alpha, \beta)}{\mu|\nu|} t^{-1-\nu} (t - r)^{\mu} + Cm_S t^{-1}. \tag{3.24}$$

For the proof of this result, we will rely on the decomposition $\partial^I L^J h_{\alpha\beta} = \sum_{k=1}^5 h_{\alpha\beta}^{IJ,k}$ with

$$\Box h_{\alpha\beta}^{IJ,1} = S_{\mathrm{Int},\alpha\beta}^{W,I,J}, \qquad h_{\alpha\beta}^{IJ,1}(2,\cdot) = 0, \quad \partial_t h_{\alpha\beta}^{IJ,1}(2,\cdot) = 0, \tag{3.25a}$$

$$\Box h_{\alpha\beta}^{IJ,2} = S_{\alpha\beta}^{KG,I,J}, \qquad h_{\alpha\beta}^{IJ,2}(2,\cdot) = 0, \quad \partial_t h_{\alpha\beta}^{IJ,2}(2,\cdot) = 0, \tag{3.25b}$$

$$\Box h_{\alpha\beta}^{IJ,3} = S_{\mathrm{Ext},\alpha\beta}^{W,I,J}, \qquad h_{\alpha\beta}^{IJ,3}(2,\cdot) = 0, \quad \partial_t h_{\alpha\beta}^{IJ,3}(2,\cdot) = 0, \tag{3.25c}$$

$$\Box h_{\alpha\beta}^{IJ,4} = 0, \qquad h_{\alpha\beta}^{IJ,4}(2,\cdot) = I_{\mathrm{Int},\alpha\beta}^{0,I,J}, \qquad \partial_t h_{\alpha\beta}^{IJ,4}(2,\cdot) = I_{\alpha\beta}^{1,I,J}, \tag{3.25d}$$

$$\Box h_{\alpha\beta}^{IJ,5} = 0, \qquad h_{\alpha\beta}^{IJ,5}(2,\cdot) = I_{\mathrm{Ext},\alpha\beta}^{0,I,J}, \quad \partial_t h_{\alpha\beta}^{IJ,5}(2,\cdot) = 0. \tag{3.25e}$$

The proof of Proposition 3.4 is immediate once we control each term.

First of all, the estimates for $h_{\alpha\beta}^{IJ,1}$ and $h_{\alpha\beta}^{IJ,2}$ are immediate from Proposition 3.1 in [41], since they concern compactly supported sources. The control of $h_{\alpha\beta}^{IJ,4}$ is standard for the homogeneous wave equation with *compact* initial data.

Lemma 3.7. *The metric coefficients satisfy the inequality*

$$\begin{aligned}|h_{\alpha\beta}^{IJ,4}(t,x)| \leq Ct^{-1}\Big(&\|\partial^I L^J h_{\alpha\beta}(2,\cdot)\|_{W^{1,\infty}(\{r\leq 1\})} \\ &+ \|\partial_t \partial^I L^J h_{\alpha\beta}(2,\cdot)\|_{L^\infty(\{r\leq 1\})}\Big)\mathbb{1}_{\{|t+2-r|\leq 1\}}(t,x).\end{aligned} \tag{3.26}$$

We thus need to study the behavior of $h_{\alpha\beta}^{IJ,3}$ and $h_{\alpha\beta}^{IJ,5}$. We treat first the function $h_{\alpha\beta}^{IJ,5}$ and observe that

$$\begin{aligned} &h_{\alpha\beta}^{IJ,5}(t,x) \\ &= \frac{1}{4\pi(t-2)^2}\int_{|y-x|=t-2}\Big(I_{\mathrm{Ext},\alpha\beta}^{0,I,J}(y) - \langle \nabla I_{\mathrm{Ext},\alpha\beta}^{0,I,J}(y), x-y\rangle\Big)\,d\sigma(y) \\ &= \frac{1}{4\pi(t-2)^2}\int_{|y-x|=t-2} I_{\mathrm{Ext},\alpha\beta}^{0,I,J}(y)d\sigma(y) \\ &\quad - \frac{1}{4\pi(t-2)^2}\int_{|y-x|=t-2}\langle \nabla I_{\mathrm{Ext},\alpha\beta}^{0,I,J}(y), x-y\rangle d\sigma(y). \end{aligned} \tag{3.27}$$

We now estimate the two integral terms successively.

Lemma 3.8. *One has* $\left|\int_{|y-x|=t} I_{Ext,\alpha\beta}^{0,I,J}(y)d\sigma(y)\right| \leq C m_S t$.

Proof. Since $g_{\alpha\beta}$ coincides with the Schwarzschild metric outside $\{r \geq 1\}$, we have immediately $|I^{0,I,J}_{\mathrm{Ext},\alpha\beta}| \leq Cm_S(1+r)^{-1}$ and thus

$$\left|\int_{|y-x|=t} I^{0,I,J}_{\mathrm{Ext},\alpha\beta}(y)d\sigma(y)\right| \leq Cm_S \int_{|y-x|=t} \frac{d\sigma(y)}{1+|y|} =: Cm_S\,\Theta(t,x). \quad (3.28)$$

Assume that $r > 0$ and, without loss of generality, $x = (r,0,0)$. Introduce the parametrization of the sphere $\{|y-x| = t\}$ such that:

- $\theta \in [0,\pi]$ is the angle from $(-1,0,0)$ to $y - x$.
- $\varphi \in [0,2\pi)$ is the angle from the plane determined by $(1,0,0)$ and $(0,1,0)$ to the plane determined by $y - x$ and $(1,0,0)$.

With this parametrization, $d\sigma(y) = t^2 \sin\theta d\theta d\varphi$ and the above integral reads

$$\Theta(t,x) = \int_{|y-x|=t} \frac{d\sigma(y)}{1+|y|} = t^2 \int_0^{2\pi}\int_0^{\pi} \frac{\sin\theta d\theta d\varphi}{1+t\big(1+(r/t)^2-(2r/t)\cos\theta\big)^{1/2}},$$

where the law of cosines was applied to $|y|$. Then, we have

$$\begin{aligned}\Theta(t,x) &= 2\pi t^2 \int_0^{\pi} \frac{\sin\theta d\theta}{1+t\big(1+(r/t)^2-(2r/t)\cos\theta\big)^{1/2}} \\ &= 2\pi t^2 \int_{-1}^{1} \frac{d\sigma}{1+t|1+(r/t)^2-(2r/t)\sigma|^{1/2}},\end{aligned}$$

with the change of variable $\sigma := \cos\theta$, so that $\lambda := t|1+(r/t)^2-(2r/t)\sigma|^{1/2}$ and

$$\Theta(t,x) = 2\pi t r^{-1}\int_{t-r}^{t+r} \frac{\lambda d\lambda}{1+\lambda} = 4\pi t - 2\pi t r^{-1}\ln\left(\frac{t+r+1}{t-r+1}\right).$$

The second term is bounded by the following observation. When $r \geq t/2$, this term is bounded by $\ln(t+1)$. When $r \leq t/2$, according to the mean value theorem, there exists ξ such that

$$r^{-1}\ln\left(\frac{t+r+1}{t-r+1}\right) = 2\,\frac{(\ln(1+t+r)-\ln(1+t-r))}{2r} = \frac{2}{1+t+\xi}.$$

By recalling $r \leq t/2$, we deduce that $\left|r^{-1}\ln\left(\frac{t+r+1}{t-r+1}\right)\right| \leq \frac{C}{1+t}$ and we conclude that the first term in the right-hand side of (3.28) is bounded by

$$Cm_S \int_{|y-x|=t} \frac{d\sigma(y)}{1+|y|} \leq Cm_S t.$$

We also observe that, when $r = 0$, we have $\int_{|y|=t} \frac{d\sigma(y)}{1+|y|} = \frac{4\pi t^2}{1+t}$ and thus $Cm_S \int_{|y-x|=t} \frac{d\sigma(y)}{1+|y|} \leq Cm_S t$. □

The proof of the following lemma is similar to the one above and we omit the proof.

Lemma 3.9. *One has*

$$\left| \int_{|y-x|=t} \langle \nabla I^{0,I,J}_{Ext,\alpha\beta}(y), x-y \rangle d\sigma(y) \right| \leq Cm_S t.$$

From the above two lemmas, we conclude that $\left|h^{IJ,5}_{\alpha\beta}(t,x)\right| \leq Cm_S t^{-1}$ as expected, and we can finally turn our attention to the last term $h^{IJ,3}_{\alpha\beta}$.

Lemma 3.10. *One has* $|h^{IJ,3}_{\alpha\beta}(t,x)| \leq Cm_S^2 t^{-1}$.

Proof. This estimate is based on Lemma 3.6 and on the explicit formula

$$h^{IJ,3}_{\alpha\beta}(t,x) = \frac{1}{4\pi}\int_2^t \frac{1}{t-s}\int_{|y|=t-s} S^{W,I,J}_{\text{Ext},\alpha\beta} d\sigma(y) ds,$$

which yields us

$$\begin{aligned}
|h^{IJ,3}_{\alpha\beta}(t,x)| &\leq Cm_S^2 \int_2^t \frac{1}{t-s}\int_{|y|=t-s} \frac{\mathbb{1}_{\{|x-y|\geq s-1\}} d\sigma}{|x-y|^4} ds \\
&= Cm_S^2 t^{-2} \int_{2/t}^1 \frac{1}{1-\lambda}\int_{|y|=1-\lambda} \frac{\mathbb{1}_{\{|y-x/t|\geq \lambda-1/t\}} d\sigma}{|y-x/t|^4} d\lambda
\end{aligned}$$

thanks to the change of variable $\lambda := s/t$. Without loss of generality, we set $x=(r,0,0)$ and introduce the following parametrization of the sphere $\{|y|=1-\lambda\}$:

- θ denotes the angle from $(1,0,0)$ to y.
- φ denotes the angle from the plane determined by $(1,0,0)$ and $(0,1,0)$ to the plane determined by $(1,0,0)$ and y.

We have $d\sigma(y) = (1-\lambda)^2 \sin\theta d\theta d\varphi$ and we must evaluate the integral $|h^{IJ,3}_{\alpha\beta}(t,x)|$

$$\begin{aligned}
&\leq Cm_S^2 t^{-2} \int_{2/t}^1 \frac{d\lambda}{1-\lambda}\int_0^{2\pi}\int_0^{\pi} \frac{\mathbb{1}_{\{|y-x/t|\geq \lambda-1/t\}} (1-\lambda)^2 \sin\theta d\theta d\varphi}{|(r/t)^2 + (1-\lambda)^2 - 2(r/t)(1-\lambda)\cos\theta|^2} \\
&\leq Cm_S^2 t^{-2} \int_{2/t}^1 \frac{d\lambda}{1-\lambda}\int_0^{\pi} \frac{\mathbb{1}_{\{|y-x/t|\geq \lambda-1/t\}} (1-\lambda)^2 \sin\theta d\theta}{|(r/t)^2 + (1-\lambda)^2 - 2(r/t)(1-\lambda)\cos\theta|^2}.
\end{aligned}$$

Consider the integral expression

$$\begin{aligned}
I(\lambda) :=& \int_0^{\pi} \frac{\mathbb{1}_{\{|y-x/t|\geq \lambda-1/t\}} (1-\lambda)^2 \sin\theta d\theta}{|(r/t)^2 + (1-\lambda)^2 - 2(r/t)(1-\lambda)\cos\theta|^2} \\
=& (1-\lambda) t r^{-1} \int_{|1-\lambda-r/t|}^{1-\lambda+r/t} \frac{\mathbb{1}_{\{\tau \geq \lambda-1/t\}} d\tau}{\tau^3},
\end{aligned}$$

where we used the change of variable $\tau := |(r/t)^2 + (1-\lambda)^2 - 2(r/t)(1-\lambda)\cos\theta|^{1/2}$. We see that when $1-\lambda+r/t \leq \lambda - 1/t$, $I(\lambda) = 0$. We only need to discuss the case $1-\lambda+r/t \geq \lambda - 1/t$ which is equivalent to $\lambda \leq \frac{t+r+1}{2t}$. We distinguish between the following cases:

• Case $1 \leq t-r \leq 3$. In this case, when $\lambda \in [2/t, (t+r+1)/2t]$, we observe that $|1-\lambda-r/t| \leq \lambda - 1/t$. Then, we find $I(\lambda) = (1-\lambda)tr^{-1}\int_{\lambda-1/t}^{1-\lambda+r/t} \frac{\mathbb{1}_{\{\tau\geq\lambda-1/t\}}d\tau}{\tau^3}$, which leads us to

$$I(\lambda) = (1-\lambda)tr^{-1}\int_{\lambda-1/t}^{1-\lambda+r/t}\frac{d\tau}{\tau^3} = \frac{t(1-\lambda)}{2r}\left((\lambda-1/t)^{-2} - (1-\lambda+r/t)^{-2}\right).$$

Then we conclude that

$$\begin{aligned}
|h_{\alpha\beta}^{IJ,3}(t,x)| &\leq Cm_S^2 t^{-2}\int_{2/t}^{(t+r+1)/2t}(1-\lambda)^{-1}I(\lambda)d\lambda \\
&= Cm_S^2 r^{-1}t^{-1}\int_{2/t}^{(t+r+1)/2t}\left((\lambda-1/t)^{-2} - (1-\lambda+r/t)^{-2}\right)d\lambda \\
&= Cm_S^2 r^{-1}\left(1 - \frac{1}{t+r-2}\right) \leq Cm_S^2 t^{-1}.
\end{aligned}$$

• Case $t-r > 3$ and $\frac{t-r}{t} \leq \frac{t+r+1}{2t} \Leftrightarrow r \geq \frac{t-1}{3}$. In this case the interval $\left[2/t, \frac{t+r+1}{2t}\right]$ is divided into two parts: $\left[2/t, \frac{t-r}{t}\right] \cup \left[\frac{t-r}{t}, \frac{t+r+1}{2t}\right]$. In the first subinterval, $|1-\lambda-r/t| = 1-\lambda-r/t$ while in the second $|1-\lambda-r/t| = r/t-1+\lambda$

Again in the subinterval $\left[2/t, \frac{t-r}{t}\right]$, we see that when $2/t \leq \lambda \leq \frac{t-r+1}{2t}$, $\lambda - 1/t \leq 1-\lambda-r/t$, when $\frac{t-r+1}{2t} \leq \lambda \leq \frac{t-r}{t}$, $\lambda - 1/t \geq 1-\lambda-r/t$. In the subinterval $\left[\frac{t-r}{t}, \frac{t+r+1}{2t}\right]$, we see that $\lambda - 1/t \geq r/t - 1 + \lambda$.

Case 1. When $\lambda \in \left[2/t, \frac{t-r+1}{2t}\right]$, we have

$$I(\lambda) = (1-\lambda)tr^{-1}\int_{1-\lambda-r/t}^{1-\lambda+r/t}\frac{d\tau}{\tau^3} = \frac{2(1-\lambda)^2}{\left((1-\lambda)^2 - (r/t)^2\right)^2}.$$

Case 2. When $\lambda \in \left[\frac{t-r+1}{2t}, \frac{t-r}{t}\right]$, we have

$$I(\lambda) = (1-\lambda)tr^{-1}\int_{\lambda-1/t}^{1-\lambda+r/t}\frac{d\tau}{\tau^3} = \frac{t(1-\lambda)}{2r}\left((\lambda-1/t)^{-2} - (1-\lambda+r/t)^{-2}\right).$$

Case 3. When $\lambda \in \left[\frac{t-r}{t}, \frac{t+r+1}{2t}\right]$, we have

$$I(\lambda) = (1-\lambda)tr^{-1}\int_{\lambda-1/t}^{1-\lambda+r/t}\frac{d\tau}{\tau^3} = \frac{t(1-\lambda)}{2r}\left((\lambda-1/t)^{-2} - (1-\lambda+r/t)^{-2}\right).$$

We obtain

$$
\begin{aligned}
|h^{IJ,3}_{\alpha\beta}(t,x)| &\leq C m_S^2 t^{-2} \int_{2/t}^{(t+r+1)/2t} (1-\lambda)^{-1} I(\lambda) d\lambda \\
&= C m_S^2 t^{-2} \int_{2/t}^{\frac{t-r+1}{2t}} + \int_{\frac{t-r+1}{2t}}^{\frac{t+r+1}{2t}} (1-\lambda)^{-1} I(\lambda) d\lambda \\
&= C m_S^2 t^{-2} \int_{2/t}^{\frac{t-r+1}{2t}} \frac{2(1-\lambda)}{\big((1-\lambda)^2 - (r/t)^2\big)^2} d\lambda \\
&\quad + C m_S^2 r^{-1} t^{-1} \int_{\frac{t-r+1}{2t}}^{\frac{t+r+1}{2t}} \big((\lambda - 1/t)^{-2} - (1-\lambda + r/t)^{-2}\big) d\lambda
\end{aligned}
$$

and we observe that

$$
\begin{aligned}
&\int_{2/t}^{\frac{t-r+1}{2t}} \frac{(1-\lambda) d\lambda}{\big((1-\lambda)^2 - (r/t)^2\big)^2} \\
&= \frac{2t^2}{(t-r-1)(t+3r-1)} - \frac{t^2}{2(t-r-2)(t+r-2)} \simeq Ct
\end{aligned}
$$

and

$$
\begin{aligned}
&\int_{\frac{t-r+1}{2t}}^{\frac{t+r+1}{2t}} \big((\lambda - 1/t)^{-2} - (1-\lambda + r/t)^{-2}\big) d\lambda \\
&= \frac{4rt}{(t-r-1)(t+r-1)} - \frac{4tr}{(t+r-1)(t+3r-1)} \simeq Cr.
\end{aligned}
$$

We conclude that $|h^{IJ,3}_{\alpha\beta}(t,x)| \leq C m_S^2 t^{-1}$.

• Case $1 - r/t \geq \frac{t+r+1}{2t} \Leftrightarrow r \leq \frac{t-1}{3}$. In this case, for $\lambda \in \big[2/t, \frac{t+r+1}{2t}\big]$, $|1-\lambda-r/t| = 1-\lambda-r/t$. We also observe that when $2/t < \lambda < \frac{t-r+1}{2t}$, $|1-\lambda-r/t| \geq \lambda - 1/t$ and when $\frac{t-r+1}{2t} \leq \lambda \leq \frac{t+r+1}{2t}$, $|1-\lambda-r/t| \leq \lambda - 1/t$. So, similarly to the above case, we find

$$
\begin{aligned}
|h^{IJ,3}_{\alpha\beta}(t,x)| &\leq C m_S^2 t^{-2} \int_{2/t}^{(t+r+1)/2t} (1-\lambda)^{-1} I(\lambda) d\lambda \\
&= C m_S^2 t^{-2} \int_{2/t}^{\frac{t-r+1}{2t}} + \int_{\frac{t-r+1}{2t}}^{\frac{t+r+1}{2t}} (1-\lambda)^{-1} I(\lambda) d\lambda \\
&= C m_S^2 t^{-2} \int_{2/t}^{\frac{t-r+1}{2t}} \frac{(1-\lambda)}{\big((1-\lambda)^2 - (r/t)^2\big)^2} d\lambda \\
&\quad + C m_S^2 r^{-1} t^{-1} \int_{\frac{t-r+1}{2t}}^{\frac{t+r+1}{2t}} \big((\lambda - 1/t)^{-2} - (1-\lambda + r/t)^{-2}\big) d\lambda,
\end{aligned}
$$

$$\int_{2/t}^{\frac{t-r+1}{2t}} \frac{(1-\lambda)d\lambda}{\left((1-\lambda)^2-(r/t)^2\right)^2}$$
$$= \frac{2t^2}{(t-r-1)(t+3r-1)} - \frac{t^2}{2(t-r-2)(t+r-2)} \simeq C$$

and

$$\int_{\frac{t-r+1}{2t}}^{\frac{t+r+1}{2t}} \left((\lambda-1/t)^{-2}-(1-\lambda+r/t)^{-2}\right) d\lambda$$
$$= \frac{4rt}{(t-r-1)(t+r-1)} - \frac{4tr}{(t+r-1)(t+3r-1)} \simeq C.$$

So, we obtain $|h_{\alpha\beta}^{IJ,3}(t,x)| \leq Cm_S^2 t^{-1}$, which completes the proof. □

3.4 Sup-norm estimate for Klein-Gordon equations

Our next statement, first presented in [41], was motivated by a pioneering work by Klainerman [33] for Klein-Gordon equations. In more recent years, Katayama [30; 31] also made some important contribution on the global existence problem for Klein-Gordon equations. Furthermore, a related estimate in two spatial dimensions in Minkowski spacetime was established earlier by Delort et al. [17]. (Our approach below could also be applied [47] in 2 + 1 dimensions.)

For compact Schwarzschild perturbations, the scalar field ϕ is supported in $\mathcal{K}$, and the sup-norm estimate in [41] remains valid for our purpose and we only need to state the corresponding result. Namely, let us consider the Klein-Gordon problem on a curved spacetime

$$-\widetilde{\Box}_g v + c^2 v = f, \qquad v|_{\mathcal{H}_2} = v_0, \quad \partial_t v|_{\mathcal{H}_2} = v_1, \tag{3.29}$$

with initial data v_0, v_1 which are prescribed on the hyperboloid $\mathcal{H}_2$ and are assumed to be compactly supported in $\mathcal{H}_2 \cap \mathcal{K}$, while the curved metric has the form $g^{\alpha\beta} = m^{\alpha\beta} + h^{\alpha\beta}$ with $\sup |\overline{h}^{00}| \leq 1/3$.

We consider the coefficient $\overline{h}^{00}$ along lines from the origin and, more precisely, we set

$$h_{t,x}(\lambda) := \overline{h}^{00}\Big(\lambda\frac{t}{s}, \lambda\frac{x}{s}\Big), \qquad s = \sqrt{t^2-r^2},$$

while $h'_{t,x}(\lambda)$ stands for the derivative with respect to the variable λ. We also set

$$s_0 := \begin{cases} 2, & 0 \leq r/t \leq 3/5, \\ \sqrt{\dfrac{t+r}{t-r}}, & 3/5 \leq r/t \leq 1. \end{cases}$$

Fixing some constant $C > 0$, we introduce the following function V by distinguishing between the regions "near" and "far" from the light cone:

$$V := \begin{cases} \Big(\|v_0\|_{L^\infty(\mathcal{H}_2)} + \|v_1\|_{L^\infty(\mathcal{H}_2)}\Big)\Big(1 + \int_2^s |h'_{t,x}(\overline{s})| e^{C\int_{\overline{s}}^s |h'_{t,x}(\lambda)|d\lambda}\, d\overline{s}\Big) & \\ \qquad + F(s) + \int_2^s F(\overline{s})|h'_{t,x}(\overline{s})| e^{C\int_{\overline{s}}^s |h'_{t,x}(\lambda)|d\lambda}\, d\overline{s}, & 0 \le r/t \le 3/5, \\ F(s) + \int_{s_0}^s F(\overline{s})|h'_{t,x}(\overline{s})| e^{C\int_{\overline{s}}^s |h'_{t,x}(\lambda)|d\lambda}\, d\overline{s}, & 3/5 < r/t < 1, \end{cases}$$

where the function F takes the right-hand side of the Klein-Gordon equation into account, as well as the curved part of the metric (except the $\overline{h}^{00}$ contribution), that is,

$$F(\overline{s}) := \int_{s_0}^{\overline{s}} \Big(|R_1[v]| + |R_2[v]| + |R_3[v]| + \lambda^{3/2}|f|\Big)(\lambda t/s, \lambda x/s)\, d\lambda$$

with

$$\begin{aligned} R_1[v] &= s^{3/2}\sum_a \overline{\partial}_a\overline{\partial}_a v + \frac{x^a x^b}{s^{1/2}}\overline{\partial}_a\overline{\partial}_b v + \frac{3}{4s^{1/2}}v + \sum_a \frac{3x^a}{s^{1/2}}\overline{\partial}_a v, \\ R_2[v] &= \overline{h}^{00}\Big(\frac{3v}{4s^{1/2}} + 3s^{1/2}\overline{\partial}_0 v\Big) \\ &\quad + s^{3/2}\big(2\overline{h}^{0b}\overline{\partial}_0\overline{\partial}_b v + \overline{h}^{ab}\overline{\partial}_a\overline{\partial}_b v + h^{\alpha\beta}\partial_\alpha\overline{\Psi}_\beta^{\beta'}\,\overline{\partial}_{\beta'} v\big), \\ R_3[v] &= \overline{h}^{00}\Big(2x^a s^{1/2}\overline{\partial}_0\overline{\partial}_a v + \frac{2x^a}{s^{1/2}}\overline{\partial}_a v + \frac{x^a x^b}{s^{1/2}}\overline{\partial}_a\overline{\partial}_b v\Big). \end{aligned}$$

Proposition 3.5. (A sup-norm estimate for Klein-Gordon equations on a curved spacetime). *Spatially compact solutions v to the Klein-Gordon problem (3.29) defined the region $\mathcal{K}_{[2,+\infty)}$ satisfy the decay estimate (for all relevant (t,x))*

$$s^{3/2}|v(t,x)| + (s/t)^{-1}s^{3/2}|\underline{\partial}_\perp\, v(t,x)| \le C\,V(t,x). \tag{3.30}$$

We postpone the proof to the Appendix.

3.5 Weighted Hardy inequality along the hyperboloidal foliation

We now derive a modified version of the Hardy inequality, formulated on hyperboloids, which is nothing but a weighted version of Proposition 5.3.1 in [39]. This inequality will play an essential role in our derivation of a key L^2 estimate for the metric component $\underline{h}^{00}$. (Cf. Section 7.2, below.)

Proposition 3.6 (Weighted Hardy inequality on hyperboloids). *For every smooth function u supported in the cone $\mathcal{K}$, one has (for any given $0 \leq \sigma \leq 1$):*

$$
\begin{aligned}
&\|(s/t)^{-\sigma} s^{-1} u\|_{L^2_f(\mathcal{H}_s)} \\
&\leq C\|(s_0/t)^{-\sigma} s_0^{-1} u\|_{L^2(\mathcal{H}_{s_0})} + C\sum_a \|\underline{\partial}_a u\|_{L^2_f(\mathcal{H}_s)} \\
&\quad + C\sum_a \int_{s_0}^{s} \tau^{-1}\Big(\|(s/t)^{1-\sigma}\partial_a u\|_{L^2(\mathcal{H}_\tau)} + \|\underline{\partial}_a u\|_{L^2(\mathcal{H}_\tau)}\Big)\, d\tau.
\end{aligned} \tag{3.31}
$$

The proof is similar to that of Proposition 5.3.1 in [39] (but we must now cope with the parameter σ) and uses the following inequality, established in [39, Chapter 5, Lemma 5.3.1].

Lemma 3.11. *For all (sufficiently regular) functions u supported in the cone $\mathcal{K}$, one has*

$$
\|r^{-1} u\|_{L^2_f(\mathcal{H}_s)} \leq C\sum_a \|\underline{\partial}_a u\|_{L^2_f(\mathcal{H}_s)}. \tag{3.32}
$$

Proof of Proposition 3.6.
Consider the vector field $W := \Big(0, -(s/t)^{-2\sigma}\frac{x^a t u^2 \chi(r/t)^2}{(1+r^2)s^2}\Big)$ defined on $\mathbb{R}^4$ and, similarly to what we did in the proof of Proposition 5.3.1 in [39], let us calculate its divergence:

$$
\begin{aligned}
\text{div}\, W = & -2s^{-1}(s/t)^{-\sigma}\sum_a \partial_a u (s/t)^{-\sigma}\frac{r\chi(r/t)u}{(1+r^2)^{1/2}s}\frac{x^a t\chi(r/t)}{r(1+r^2)^{1/2}} \\
& - 2s^{-1}(s/t)^{-\sigma} r^{-1} u (s/t)^{-\sigma}\frac{r\chi(r/t)u}{(1+r^2)^{1/2}s}\frac{\chi'(r/t)r}{(1+r^2)^{1/2}} \\
& - (s/t)^{-2\sigma}\big(u\chi(r/t)\big)^2\left(\frac{r^2 t + 3t}{(1+r^2)^2 s^2} + \frac{2r^2 t}{(1+r^2)s^4}\right) \\
& - 2\sigma(s/t)^{-1-2\sigma}\big(u\chi(r/t)\big)^2\frac{r^2}{(1+r^2)s^3}.
\end{aligned}
$$

We integrate this identity within $\mathcal{K}_{[s_0,s_1]}$ and, after recalling the relation

$dxdt = (s/t)\,dxds$, we obtain that $\int_{\mathcal{K}_{[s_0,s_1]}} \operatorname{div} W dxdt$ equals

$$
\begin{aligned}
&-2\int_{\mathcal{K}_{[s_0,s_1]}} s^{-1}(s/t)^{1-\sigma}\sum_a \partial_a u (s/t)^{-\sigma}\frac{r\chi(r/t)u}{(1+r^2)^{1/2}s}\frac{x^a t\chi(r/t)}{r(1+r^2)^{1/2}}\,dxds\\
&-2\int_{\mathcal{K}_{[s_0,s_1]}} s^{-1}(s/t)^{1-\sigma}r^{-1}u(s/t)^{-\sigma}\frac{r\chi(r/t)u}{(1+r^2)^{1/2}s}\frac{\chi'(r/t)r}{(1+r^2)^{1/2}}\,dxds\\
&-\int_{\mathcal{K}_{[s_0,s_1]}} (s/t)^{1-2\sigma}\big(u\chi(r/t)\big)^2\left(\frac{r^2t+3t}{(1+r^2)^2s^2}+\frac{2r^2t}{(1+r^2)s^4}\right)dxds\\
&-2\sigma\int_{\mathcal{K}_{[s_0,s_1]}} (s/t)^{-2\sigma}\big(u\chi(r/t)\big)^2\frac{r^2}{(1+r^2)s^3}\,dxds,
\end{aligned}
$$

thus

$$
\begin{aligned}
&= -2\int_{s_0}^{s_1} ds\int_{\mathcal{H}_s} s^{-1}(s/t)^{1-\sigma}\sum_a \partial_a u(s/t)^{-\sigma}\frac{r\chi(r/t)u}{(1+r^2)^{1/2}s}\frac{x^a t\chi(r/t)}{r(1+r^2)^{1/2}}\,dx\\
&\quad -2\int_{s_0}^{s_1} ds\int_{\mathcal{H}_s} s^{-1}(s/t)^{1-\sigma}r^{-1}u(s/t)^{-\sigma}\frac{r\chi(r/t)u}{(1+r^2)^{1/2}s}\frac{\chi'(r/t)r}{(1+r^2)^{1/2}}\,dx\\
&\quad -\int_{s_0}^{s_1} ds\int_{\mathcal{H}_s} (s/t)^{1-2\sigma}\big(u\chi(r/t)\big)^2\left(\frac{r^2t+3t}{(1+r^2)^2s^2}+\frac{2r^2t}{(1+r^2)s^4}\right)dx\\
&\quad -2\sigma\int_{s_0}^{s_1} ds\int_{\mathcal{H}_s} (s/t)^{-2\sigma}\big(u\chi(r/t)\big)^2\frac{r^2}{(1+r^2)s^3}\,dx\\
&=: \int_{s_0}^{s_1}(T_1+T_2+T_3+T_4)ds.
\end{aligned}
$$

On the other hand, we apply Stokes' formula to the left-hand side of this identity. Recall that the flux vector vanishes in a neighborhood of the boundary of $\mathcal{K}_{[s_0,s_1]}$, which is $\{r=t-1, s_0\leq\sqrt{t^2-r^2}\leq s_1\}$ and, by a calculation similar to the one in the proof of Lemma 3.1,

$$
\begin{aligned}
&\left\|(s/t)^{-\sigma}\frac{r\chi(r/t)u}{(1+r^2)^{1/2}s}\right\|^2_{L^2(\mathcal{H}_{s_1})}-\left\|(s/t)^{-\sigma}\frac{r\chi(r/t)u}{(1+r^2)^{1/2}s}\right\|^2_{L^2(\mathcal{H}_{s_0})}\\
&=\int_{s_0}^{s_1}(T_1+T_2+T_3+T_4)ds.
\end{aligned}
$$

After differentiation with respect to s, we obtain

$$
\begin{aligned}
&2\left\|(s/t)^{-\sigma}\frac{r\chi(r/t)u}{(1+r^2)^{1/2}s}\right\|_{L^2(\mathcal{H}_{s_1})}\frac{d}{ds}\left\|(s/t)^{-\sigma}\frac{r\chi(r/t)u}{(1+r^2)^{1/2}s}\right\|_{L^2(\mathcal{H}_{s_1})}\\
&=T_1+T_2+T_3+T_4.
\end{aligned}
\tag{3.33}
$$

We observe that

$$
\begin{aligned}
|T_1| &\le 2\sum_a \int_{\mathcal{H}_s} s^{-1}(s/t)^{1-\sigma}|\partial_a u|(s/t)^{-\sigma}\frac{r\chi(r/t)|u|}{(1+r^2)^{1/2}s}\frac{|x^a|t\chi(r/t)}{r(1+r^2)^{1/2}}\,dx \\
&\le 2\sum_a s^{-1}\|(s/t)^{1-\sigma}\partial_a u\|_{L^2_f(\mathcal{H}_s)}\left\|(s/t)^{-\sigma}\frac{r\chi(r/t)u}{(1+r^2)^{1/2}s}\right\|_{L^2_f(\mathcal{H}_s)}\left\|\frac{x^a t\chi(r/t)}{r(1+r^2)^{1/2}}\right\|_{L^\infty(\mathcal{H}_s)} \\
&\le Cs^{-1}\sum_a \|(s/t)^{1-\sigma}\partial_a u\|_{L^2_f(\mathcal{H}_s)}\left\|(s/t)^{-\sigma}\frac{r\chi(r/t)u}{(1+r^2)^{1/2}s}\right\|_{L^2_f(\mathcal{H}_s)},
\end{aligned}
$$

where we have observed that $\left\|\frac{x^a t\chi(r/t)}{r(1+r^2)^{1/2}}\right\|_{L^\infty(\mathcal{H}_s)} \le C$, since the support of $\chi(\cdot)$ is contained in $\{r \ge t/3\}$. Similarly, we find

$$
\begin{aligned}
|T_2| &\le Cs^{-1}\|(s/t)^{1-\sigma}r^{-1}u\|_{L^2_f(\mathcal{H}_s)}\left\|(s/t)^{-\sigma}\frac{r\chi(r/t)u}{(1+r^2)^{1/2}s}\right\|_{L^2_f(\mathcal{H}_s)} \\
&\le Cs^{-1}\|r^{-1}u\|_{L^2_f(\mathcal{H}_s)}\left\|(s/t)^{-\sigma}\frac{r\chi(r/t)u}{(1+r^2)^{1/2}s}\right\|_{L^2_f(\mathcal{H}_s)} \\
&\le Cs^{-1}\sum_a \|\underline{\partial}_a u\|_{L^2_f(\mathcal{H}_s)}\left\|(s/t)^{-\sigma}\frac{r\chi(r/t)u}{(1+r^2)^{1/2}s}\right\|_{L^2_f(\mathcal{H}_s)},
\end{aligned}
$$

where we have applied (3.32). We also observe that $T_3 \le 0$ and $T_4 \le 0$. Then, (3.33) leads us to

$$
\begin{aligned}
&\frac{d}{ds}\left\|(s/t)^{-\sigma}\frac{r\chi(r/t)u}{(1+r^2)^{1/2}s}\right\|_{L^2(\mathcal{H}_{s_1})} \\
&\le Cs^{-1}\sum_a \left(\|(s/t)^{1-\sigma}\partial_a u\|_{L^2_f(\mathcal{H}_s)} + \|\underline{\partial}_a u\|_{L^2_f(\mathcal{H}_s)}\right).
\end{aligned}
\tag{3.34}
$$

Then by integrating on the interval $[s_0, s]$, we have

$$
\begin{aligned}
&\left\|(s/t)^{-\sigma}\frac{r\chi(r/t)u}{(1+r^2)^{1/2}s}\right\|_{L^2_f(\mathcal{H}_s)} \\
&\le \left\|(s/t)^{-\sigma}\frac{r\chi(r/t)u}{(1+r^2)^{1/2}s}\right\|_{L^2(\mathcal{H}_{s_0})} \\
&\quad + C\sum_a \int_{s_0}^{s} \tau^{-1}\left(\|(s/t)^{1-\sigma}\partial_a u\|_{L^2(\mathcal{H}_\tau)} + \|\underline{\partial}_a u\|_{L^2(\mathcal{H}_\tau)}\right)d\tau,
\end{aligned}
\tag{3.35}
$$

which is the desired estimate in the outer part of $\mathcal{H}_s$.

For the inner part, $r \le t/3$ leads us to $\frac{2\sqrt{2}}{3} \le s/t \le 1$. Then by Lemma 3.11, we find

$$
\left\|(s/t)^{-\sigma}\frac{r(1-\chi(r/t))u}{(1+r^2)^{1/2}s}\right\|_{L^2_f(\mathcal{H}_s)} \le \|r^{-1}u\|_{L^2_f(\mathcal{H}_s)} \le C\sum_a \|\underline{\partial}_a u\|_{L^2_f(\mathcal{H}_s)}
\tag{3.36}
$$

and it remains to combine (3.35) and (3.36). □

3.6 Sobolev inequality on hyperboloids

In order to turn an L^2 energy estimate into an L^∞ estimate, we will rely on the following version of the Sobolev inequality (Klainerman [33], Hörmander [27, Lemma 7.6.1]; see also LeFloch and Ma [39, Section 5]).

Proposition 3.7 (Sobolev-type estimate on hyperboloids). *For any sufficiently smooth function $u = u(t,x)$ which is defined in the future of $\mathcal{H}_2$ and is spatially compactly supported, one has*

$$\sup_{(t,x)\in\mathcal{H}_s} t^{3/2}\,|u(t,x)| \lesssim \sum_{|I|\le 2} \|L^I u(t,\cdot)\|_{L^2_f(\mathcal{H}_s)}, \qquad s \geq 2, \tag{3.37}$$

where the implied constant is uniform in $s \geq 2$, and one recalls that $t = \sqrt{s^2+|x|^2}$ on $\mathcal{H}_s$.

Proof. Consider the function $w_s(x) := u(\sqrt{s^2+|x|^2}, x)$. Fix s_0 and a point (t_0, x_0) in $\mathcal{H}_{s_0}$ (with $t_0 = \sqrt{s_0^2+|x_0|^2}$), and observe that

$$\partial_a w_{s_0}(x) = \underline{\partial}_a u\big(\sqrt{s_0^2+|x|^2}, x\big) = \underline{\partial}_a u(t,x), \tag{3.38}$$

with $t = \sqrt{s_0^2+|x|^2}$ and $t\partial_a w_{s_0}(x) = t\underline{\partial}_a u\big(\sqrt{s_0^2+|x|^2}, t\big) = L_a u(t,x)$. Then, introduce $g_{s_0,t_0}(y) := w_{s_0}(x_0 + t_0\, y)$ and write

$$g_{s_0,t_0}(0) = w_{s_0}(x_0) = u\big(\sqrt{s_0^2+|x_0|^2}, x_0\big) = u(t_0,x_0).$$

From the standard Sobolev inequality applied to the function g_{s_0,t_0}, we get

$$\big|g_{s_0,t_0}(0)\big|^2 \leq C \sum_{|I|\le 2} \int_{B(0,1/3)} |\partial^I g_{s_0,t_0}(y)|^2\,dy,$$

$B(0,1/3) \subset \mathbb{R}^3$ being the ball centered at the origin with radius $1/3$.

In view of (with $x = x_0 + t_0 y$)

$$\begin{aligned}\partial_a g_{s_0,t_0}(y) &= t_0 \partial_a w_{s_0}(x_0 + t_0 y)\\ &= t_0\partial_a w_{s_0}(x) = t_0 \underline{\partial}_a u(t,x)\end{aligned}$$

in view of (3.38), we have (for all I) $\partial^I g_{s_0,t_0}(y) = (t_0\underline{\partial})^I u(t,x)$ and, therefore,

$$\begin{aligned}\big|g_{s_0,t_0}(0)\big|^2 \leq& C \sum_{|I|\le 2} \int_{B(0,1/3)} \big|(t_0\underline{\partial})^I u(t,x)\big|^2 dy\\ =& C t_0^{-3} \sum_{|I|\le 2} \int_{B((t_0,x_0),t_0/3)\cap\mathcal{H}_{s_0}} \big|(t_0\underline{\partial})^I u(t,x)\big|^2 dx.\end{aligned}$$

Note that

$$\begin{aligned}(t_0\underline{\partial}_a(t_0\underline{\partial}_b w_{s_0})) &= t_0^2\underline{\partial}_a\underline{\partial}_b w_{s_0}\\ &= (t_0/t)^2(t\underline{\partial}_a)(t\underline{\partial}_b)w_{s_0} - (t_0/t)^2(x^a/t)L_b w_{s_0}\end{aligned}$$

and $x^a/t = x_0^a/t + yt_0/t = (x_0^a/t_0 + y)(t_0/t)$. In the region $y \in B(0,1/3)$, the factor $|x^a/t|$ is bounded by $C(t_0/t)$ and thus (for $|I| \leq 2$)

$$|(t_0\underline{\partial})^I u| \leq \sum_{|J|\leq|I|} |L^J u|(t_0/t)^2.$$

In the region $|x_0| \leq t_0/2$, we have $t_0 \leq \frac{2}{\sqrt{3}}s_0$ so

$$t_0 \leq Cs_0 \leq C\sqrt{|x|^2 + s_0^2} = Ct$$

for some $C > 0$. When $|x_0| \geq t_0/2$, in the region $B((t_0, x_0), t_0/3) \cap \mathcal{H}_{s_0}$ we get $t_0 \leq C|x| \leq C\sqrt{|x|^2 + s_0^2} = Ct$ and thus

$$|(t_0\underline{\partial})^I u| \leq C \sum_{|J|\leq|I|} |L^J u|$$

and

$$\begin{aligned}\left|g_{s_0,t_0}(y_0)\right|^2 \leq& Ct_0^{-3} \sum_{|I|\leq 2} \int_{B(x_0,t_0/3)\cap\mathcal{H}_{s_0}} \left|(t\underline{\partial})^I u(t,x)\right|^2 dx\\ \leq& Ct_0^{-3} \sum_{|I|\leq 2} \int_{\mathcal{H}_{s_0}} \left|L^I u(t,x)\right|^2 dx.\end{aligned}$$

□

3.7 Hardy inequality for hyperboloids

We now bound the norm $\|r^{-1}\partial^I L^J h_{\alpha\beta}\|_{L^2(\mathcal{H}_s^*)}$. If $\partial^I L^J h_{\alpha\beta}$ were compactly supported in $\mathcal{H}_s \cap \mathcal{K}$, we could directly apply the standard Hardy inequality to the function $u_s(x) := \big(\partial^I L^J h_{\alpha\beta}\big)(\sqrt{s^2+r^2}, x)$ and we would obtain

$$\|r^{-1}\partial^I L^J h_{\alpha\beta}\|_{L^2(\mathcal{H}_s^*)} \leq C\|\partial\partial^I L^J h_{\alpha\beta}\|_{L^2(\mathcal{H}_s^*)}.$$

However, since $\partial^I L^J h_{\alpha\beta}$ is not compactly supported in $\mathcal{K}$, we must take a boundary term into account.

Lemma 3.12 (Adapted Hardy inequality). *Let $(h_{\alpha\beta}, \phi)$ be a solution to the Einstein-massive field system associated with a compact Schwarzschild perturbation. Then, one has*

$$\|r^{-1}\partial^I L^J h_{\alpha\beta}\|_{L^2(\mathcal{H}_s^*)} \leq C\sum_a \|\underline{\partial}_a\partial^I L^J h_{\alpha\beta}\|_{L^2(\mathcal{H}_s^*)} + Cm_S s^{-1}. \quad (3.39)$$

Proof. With the notation $u_s(x) := (\partial^I L^J h_{\alpha\beta})(\sqrt{s^2+r^2}, x)$, we obtain

$$\partial_a u_s(x) = \underline{\partial}_a \partial^I L^J h_{\alpha\beta}\left(\sqrt{s^2+r^2}, x\right).$$

Consider the identity $r^{-2}u_s^{-2} = -\partial_r\left(r^{-1}u_s^2\right) + 2u_s r^{-1}\partial_r u_s$ and integrate it in the region $C_{[\varepsilon,(s^2-1)/2]} := \left\{\varepsilon \leq r \leq \frac{s^2-1}{2}\right\}$ with spherical coordinates. We have

$$\begin{aligned}
&\int_{C_{[\varepsilon,(s^2-1)/2]}} |r^{-1}u_s|^2 dx \\
&= \int_{r=(s^2-1)/2} r^{-1}u_s^2 d\sigma - \int_{r=\varepsilon} r^{-1}u_s^2 d\sigma + 2\int_{C_{[\varepsilon,(s^2-1)/2]}} u_s r^{-1}\partial_r u_s dx.
\end{aligned} \tag{3.40}$$

Letting now $\varepsilon \to 0^+$, we have $\int_{r=\varepsilon} r^{-1}u_s^2 d\sigma \to 0$. Observe that on the sphere $r = (s^2-1)/2$,

$$\sqrt{s^2+r^2} - r = \frac{s^2+1}{2} - \frac{s^2-1}{2} = 1,$$

that is the point $\left(\sqrt{s^2+r^2}, x\right)$ is on the cone $\{r = t-1\}$. We know that, on this cone, $h_{\alpha\beta}$ coincides with the Schwarzschild metric, so that

$$\int_{r=(s^2-1)/2} r^{-1}u_s^2 d\sigma \leq C m_S^2 s^{-2}.$$

Then, (3.40) yields us

$$\begin{aligned}
&\|r^{-1}u_s\|^2_{L^2(C_{[0,(s^2-1)/2]})} \\
&\quad < 2\|r^{-1}u_s\|_{L^2(C_{[0,(s^2-1)/2]})}\|\partial_r u_s\|_{L^2(C_{[0,(s^2-1)/2]})} + C m_S^2 s^{-2}.
\end{aligned}$$

And this inequality leads us to

$$\|r^{-1}u_s\|_{L^2(C_{[0,(s^2-1)/2]})} \leq C\|\partial_r u_s\|_{L^2(C_{[0,(s^2-1)/2]})} + C m_S s^{-1}.$$

By recalling that

$$\begin{aligned}
\|r^{-1}u_s\|^2_{L^2(C_{[0,(s^2-1)/2]})} &= \int_{r\leq(s^2-1)/2} \left|r^{-1}\partial^I L^J h_{\alpha\beta}\left(\sqrt{s^2+r^2}, x\right)\right|^2 dx \\
&= \int_{\mathcal{K}\cap\mathcal{H}_s} \left|r^{-1}\partial^I L^J h_{\alpha\beta}(t,x)\right|^2 dx = \|r^{-1}\partial^I L^J h_{\alpha\beta}\|^2_{L^2(\mathcal{H}_s^*)}
\end{aligned}$$

and $\partial_r u_s = \frac{x^a}{r}\partial_a u_s = \frac{x^a}{r}\underline{\partial}_a \partial^I L^J h_{\alpha\beta}(\sqrt{s^2+r^2}, x)$, the proof is completed. □

3.8 Commutator estimates for admissible vector fields

We recall the following identities first established in [41]; see also Appendix D at the end of this monograph.

Lemma 3.13 (Algebraic decomposition of commutators). *One has*

$$[\partial_t, \underline{\partial}_a] = -\frac{x^a}{t^2}\partial_t, \quad [\underline{\partial}_a, \underline{\partial}_b] = 0. \tag{3.41}$$

There exist constants λ_{aJ}^I such that

$$[\partial^I, L_a] = \sum_{|J|\leq|I|} \lambda_{aJ}^I \partial^J. \tag{3.42}$$

There exist constants $\theta_{\alpha J}^{I\gamma}$ such that

$$[L^I, \partial_\alpha] = \sum_{|J|<|I|,\gamma} \theta_{\alpha J}^{I\gamma} \partial_\gamma L^J. \tag{3.43}$$

In the future light-cone $\mathcal{K}$, the following identity holds:

$$[\partial^I L^J, \underline{\partial}_\beta] = \sum_{\substack{|J'|\leq|J| \\ |I'|\leq|I|}} \underline{\theta}_{\beta I'J'}^{IJ\gamma} \partial_\gamma \partial^{I'} L^{J'}, \tag{3.44}$$

where the coefficients $\underline{\theta}_{\beta I'J'}^{IJ\gamma}$ are smooth functions and satisfy (in $\mathcal{K}$)

$$\begin{aligned} \big|\partial^{I_1} L^{J_1} \underline{\theta}_{\beta I'J'}^{IJ\gamma}\big| &\leq C(|I|,|J|,|I_1|,|J_1|)\, t^{-|I_1|}, \quad |J'|<|J|, \\ \big|\partial^{I_1} L^{J_1} \underline{\theta}_{\beta I'J'}^{IJ\gamma}\big| &\leq C(|I|,|J|,|I_1|,|J_1|)\, t^{-|I_1|-1}, \quad |I'|<|I|. \end{aligned} \tag{3.45}$$

Within the future light-cone $\mathcal{K}$, the following identity holds:

$$[L^I, \underline{\partial}_c] = \sum_{|J|<|I|} \sigma_{cJ}^{Ia} \underline{\partial}_a L^J, \tag{3.46}$$

where the coefficients σ_{cJ}^{Ia} are smooth functions and satisfy (in $\mathcal{K}$)

$$\big|\partial^{I_1} L^{J_1} \sigma_{cJ}^{Ia}\big| \leq C(|I|,|J|,|I_1|,|J_1|) t^{-|I_1|}. \tag{3.47}$$

Within the future light-cone $\mathcal{K}$, the following identity holds:

$$[\partial^I, \underline{\partial}_c] = t^{-1} \sum_{|J|\leq|I|} \rho_{cJ}^I \partial^J, \tag{3.48}$$

where the coefficients ρ_{cJ}^I are smooth functions and satisfy (in $\mathcal{K}$)

$$\big|\partial^{I_1} L^{J_1} \rho_{cJ}^I\big| \leq C(|I|,|J|,|I_1|,|J_1|) t^{-|I_1|}. \tag{3.49}$$

Lemma 3.14. *For all indices I, the function*

$$\Xi^I := (t/s)\partial^I L^J(s/t) \tag{3.50}$$

defined in the closed cone $\overline{\mathcal{K}} = \{|x| \leq t-1\}$, is smooth and all of its derivatives (of any order) are bounded in $\overline{\mathcal{K}}$. Furthermore, it is homogeneous of degree η with $\eta \leq 0$ (in the sense recalled in Definition 4.1 below).

Lemma 3.15 (Commutator estimates). *For all sufficiently smooth functions u defined in the cone $\mathcal{K}$, the following identities hold:*

$$\big|[\partial^I L^J, \partial_\alpha]u\big| \leq C(|I|,|J|) \sum_{|J'|<|J|,\beta} |\partial_\beta \partial^I L^{J'} u|, \tag{3.51}$$

$$\big|[\partial^I L^J, \underline{\partial}_c]u\big| \leq C(|I|,|J|) \sum_{\substack{|J'|<|J|,a\\ |I'|\leq|I|}} |\underline{\partial}_a \partial^{I'} L^{J'} u| + C(|I|,|J|)t^{-1} \sum_{\substack{|I|\leq|I'|\\ |J|\leq|J'|}} |\partial^{I'} L^{J'} u|. \tag{3.52}$$

$$\begin{aligned}\big|[\partial^I L^J, \underline{\partial}_\alpha]u\big| \leq\ & C(|I|,|J|)t^{-1} \sum_{\substack{\beta,|I'|<|I|\\ |J'|\leq|J|}} \big|\partial_\beta \partial^{I'} L^{J'} u\big| \\ & + C(|I|,|J|) \sum_{\substack{\beta,|I'|\leq|I|\\ |J'|<|J|}} \Big|\partial_\beta \partial^{I'} L^{J'} u\Big|,\end{aligned} \tag{3.53}$$

$$\big|[\partial^I L^J, \partial_\alpha \partial_\beta]u\big| \leq C(|I|,|J|) \sum_{\substack{\gamma,\gamma'\\ |I|\leq|I'|,|J'|<|I|}} \big|\partial_\gamma \partial_{\gamma'} \partial^{I'} L^{J'} u\big|, \tag{3.54}$$

$$\begin{aligned}&\big|[\partial^I L^J, \underline{\partial}_a \underline{\partial}_\beta]u\big| + \big|[\partial^I L^J, \underline{\partial}_\alpha \underline{\partial}_b]u\big| \\ &\leq C(|I|,|J|)\Bigg(\sum_{\substack{c,\gamma,|I'|\leq|I|\\ |J'|<|J|}} \big|\underline{\partial}_c \underline{\partial}_\gamma \partial^{I'} L^{J'} u\big| + \sum_{\substack{c,\gamma,|I'|<|I|\\ |J'|\leq|J|}} t^{-1}\big|\underline{\partial}_c \underline{\partial}_\gamma \partial^{I'} L^{J'} u\big| \\ &\qquad + \sum_{\substack{\gamma,|I'|\leq|I|\\ |J'|\leq|J|}} t^{-1}\big|\partial_\gamma \partial^{I'} L^{J'} u\big|\Bigg).\end{aligned} \tag{3.55}$$

Chapter 4

Quasi-Null Structure of the Einstein-Massive Field System on Hyperboloids

4.1 Einstein equations in wave coordinates

Our next task is to derive an explicit expression for the curvature. We set $\Gamma^\gamma := g^{\alpha\beta}\Gamma_{\alpha\beta}^\gamma = 0$ and $\Gamma_\alpha := g_{\alpha\beta}\Gamma^\beta$.

Lemma 4.1 (Ricci curvature of a 4-manifold). *In arbitrary local coordinates, one has the decomposition:*

$$R_{\alpha\beta} = -\frac{1}{2}g^{\lambda\delta}\partial_\lambda\partial_\delta g_{\alpha\beta} + \frac{1}{2}\big(\partial_\alpha\Gamma_\beta + \partial_\beta\Gamma_\alpha\big) + \frac{1}{2}F_{\alpha\beta},$$

where $F_{\alpha\beta} := P_{\alpha\beta} + Q_{\alpha\beta} + W_{\alpha\beta}$ is a sum of null terms, that is,

$$\begin{aligned}
Q_{\alpha\beta} :&= g^{\lambda\lambda'}g^{\delta\delta'}\partial_\delta g_{\alpha\lambda'}\partial_{\delta'}g_{\beta\lambda} - g^{\lambda\lambda'}g^{\delta\delta'}\big(\partial_\delta g_{\alpha\lambda'}\partial_\lambda g_{\beta\delta'} - \partial_\delta g_{\beta\delta'}\partial_\lambda g_{\alpha\lambda'}\big)\\
&+ g^{\lambda\lambda'}g^{\delta\delta'}\big(\partial_\alpha g_{\lambda'\delta'}\partial_\delta g_{\lambda\beta} - \partial_\alpha g_{\lambda\beta}\partial_\delta g_{\lambda'\delta'}\big)\\
&+ \frac{1}{2}g^{\lambda\lambda'}g^{\delta\delta'}\big(\partial_\alpha g_{\lambda\beta}\partial_{\lambda'}g_{\delta\delta'} - \partial_\alpha g_{\delta\delta'}\partial_{\lambda'}g_{\lambda\beta}\big)\\
&+ g^{\lambda\lambda'}g^{\delta\delta'}\big(\partial_\beta g_{\lambda'\delta'}\partial_\delta g_{\lambda\alpha} - \partial_\beta g_{\lambda\alpha}\partial_\delta g_{\lambda'\delta'}\big)\\
&+ \frac{1}{2}g^{\lambda\lambda'}g^{\delta\delta'}\big(\partial_\beta g_{\lambda\alpha}\partial_{\lambda'}g_{\delta\delta'} - \partial_\beta g_{\delta\delta'}\partial_{\lambda'}g_{\lambda\alpha}\big),
\end{aligned}$$

quasi-null term (as they are called by the authors)

$$P_{\alpha\beta} := -\frac{1}{2}g^{\lambda\lambda'}g^{\delta\delta'}\partial_\alpha g_{\delta\lambda'}\partial_\beta g_{\lambda\delta'} + \frac{1}{4}g^{\delta\delta'}g^{\lambda\lambda'}\partial_\beta g_{\delta\delta'}\partial_\alpha g_{\lambda\lambda'}$$

and a remainder $W_{\alpha\beta} := g^{\delta\delta'}\partial_\delta g_{\alpha\beta}\Gamma_{\delta'} - \Gamma_\alpha\Gamma_\beta$.

Let us make some observations based on this lemma. Note that the Einstein equation $R_{\alpha\beta} = 0$ now reads

$$\widetilde{\Box}_g h_{\alpha\beta} = P_{\alpha\beta} + Q_{\alpha\beta} + W_{\alpha\beta} + \big(\partial_\alpha\Gamma_\beta + \partial_\beta\Gamma_\alpha\big). \tag{4.1}$$

Furthermore, if the coordinates are assumed to satisfy the wave condition $\Gamma^\gamma = 0$, so that $\Gamma_\beta = 0$ and, by specifying the dependence of the right-hand sides in $(g; \partial h)$,

$$\widetilde{\Box}_g g_{\alpha\beta} = P_{\alpha\beta}(g; \partial h) + Q_{\alpha\beta}(g; \partial h), \tag{4.2}$$

which is a standard result.

For the Einstein-massive field system

$$\begin{aligned} G_{\alpha\beta} &= 8\pi T_{\alpha\beta}, \\ T_{\alpha\beta} &= \partial_\alpha\phi\partial_\beta\phi - \frac{1}{2}g_{\alpha\beta}\left(g^{\mu\nu}\partial_\mu\phi\partial_\nu\phi + c^2\phi^2\right), \end{aligned} \tag{4.3}$$

we obtain

$$R_{\alpha\beta} = 8\pi\left(\nabla_\alpha\phi\nabla_\beta\phi + \frac{1}{2}c^2\phi^2 g_{\alpha\beta}\right)$$

and, by the above lemma, the Einstein-massive field system in a wave coordinate system reads

$$\begin{aligned} &\widetilde{\Box}_g g_{\alpha\beta} = P_{\alpha\beta}(g; \partial h) + Q_{\alpha\beta}(g; \partial h) - 16\pi\partial_\alpha\phi\partial_\beta\phi - 8\pi c^2\phi^2 g_{\alpha\beta}, \\ &\widetilde{\Box}_g\phi - c^2\phi = 0. \end{aligned} \tag{4.4}$$

Proof of Lemma 4.1. We need to perform straightforward but very tedious calculations, starting from the definitions

$$\begin{aligned} R_{\alpha\beta} &= \partial_\lambda\Gamma^\lambda_{\alpha\beta} - \partial_\alpha\Gamma^\lambda_{\beta\lambda} + \Gamma^\lambda_{\alpha\beta}\Gamma^\delta_{\lambda\delta} - \Gamma^\lambda_{\alpha\delta}\Gamma^\delta_{\beta\lambda}, \\ \Gamma^\lambda_{\alpha\beta} &= \frac{1}{2}g^{\lambda\lambda'}\left(\partial_\alpha g_{\beta\lambda'} + \partial_\beta g_{\alpha\lambda'} - \partial_{\lambda'}g_{\alpha\beta}\right). \end{aligned}$$

Only the first two terms in the expression $R_{\alpha\beta}$ involves second-order derivatives of the metric, and we focus on those terms first. In view of

$$\begin{aligned} \partial_\lambda\Gamma^\lambda_{\alpha\beta} &= -\frac{1}{2}g^{\lambda\delta}\partial_\lambda\partial_\delta g_{\alpha\beta} + \frac{1}{2}g^{\lambda\delta}\partial_\lambda\partial_\alpha g_{\beta\delta} + \frac{1}{2}g^{\lambda\delta}\partial_\lambda\partial_\beta g_{\alpha\delta} \\ &\quad + \frac{1}{2}\partial_\lambda g^{\lambda\delta}\left(\partial_\alpha g_{\beta\delta} + \partial_\beta g_{\alpha\delta} - \partial_\delta g_{\alpha\beta}\right), \\ \partial_\alpha\Gamma^\lambda_{\beta\lambda} &= \frac{1}{2}\partial_\alpha\partial_\beta g_{\lambda\delta} + \frac{1}{2}\partial_\alpha g^{\lambda\delta}\partial_\beta g_{\lambda\delta}, \end{aligned}$$

we can write

$$\begin{aligned} &\partial_\lambda\Gamma^\lambda_{\alpha\beta} - \partial_\alpha\Gamma^\lambda_{\beta\lambda} \\ &= -\frac{1}{2}g^{\lambda\delta}\partial_\lambda\partial_\delta g_{\alpha\beta} + \frac{1}{2}g^{\lambda\delta}\partial_\alpha\partial_\lambda g_{\delta\beta} + \frac{1}{2}g^{\lambda\delta}\partial_\beta\partial_\lambda g_{\delta\alpha} - \frac{1}{2}g^{\lambda\delta}\partial_\alpha\partial_\beta g_{\lambda\delta} \\ &\quad - \frac{1}{2}\partial_\lambda g^{\lambda\delta}\partial_\delta g_{\alpha\beta} + \frac{1}{2}\partial_\lambda g^{\lambda\delta}\partial_\alpha g_{\beta\delta} + \frac{1}{2}\partial_\lambda g^{\lambda\delta}\partial_\beta g_{\alpha\delta} - \frac{1}{2}\partial_\alpha g^{\lambda\delta}\partial_\beta g_{\lambda\delta}, \end{aligned} \tag{4.5}$$

in which the first line contains second-order terms and the second line contains quadratic products of first-order terms.

Let us next compute the term $\partial_\alpha \Gamma_\beta + \partial_\beta \Gamma_\alpha$ (which appears in our decomposition). We have

$$
\begin{aligned}
\Gamma^\gamma = g^{\alpha\beta}\Gamma^\gamma_{\alpha\beta} &= \frac{1}{2} g^{\alpha\beta} g^{\gamma\delta}\big(\partial_\alpha g_{\beta\delta} + \partial_\beta g_{\alpha\delta} - \partial_\delta g_{\alpha\beta}\big) \\
&= g^{\gamma\delta} g^{\alpha\beta} \partial_\alpha g_{\beta\delta} - \frac{1}{2} g^{\alpha\beta} g^{\gamma\delta} \partial_\delta g_{\alpha\beta}
\end{aligned}
$$

and, therefore, $\Gamma_\lambda = g_{\lambda\gamma}\Gamma^\gamma = g^{\alpha\beta}\partial_\alpha g_{\beta\lambda} - \frac{1}{2} g^{\alpha\beta}\partial_\lambda g_{\alpha\beta}$, so that, after differentiating,

$$
\begin{aligned}
\partial_\alpha \Gamma_\beta &= \partial_\alpha\big(g^{\delta\lambda}\partial_\delta g_{\lambda\beta}\big) - \frac{1}{2}\partial_\alpha\big(g^{\lambda\delta}\partial_\beta g_{\lambda\delta}\big) \\
&= g^{\delta\lambda}\partial_\alpha\partial_\delta g_{\lambda\beta} - \frac{1}{2} g^{\lambda\delta}\partial_\alpha\partial_\beta g_{\lambda\delta} - \frac{1}{2}\partial_\alpha g^{\lambda\delta}\partial_\beta g_{\lambda\delta} + \partial_\alpha g^{\delta\lambda}\partial_\delta g_{\lambda\beta}.
\end{aligned}
$$

The term of interest is thus found to be

$$
\begin{aligned}
&\partial_\alpha\Gamma_\beta + \partial_\beta\Gamma_\alpha \\
&= g^{\lambda\delta}\partial_\alpha\partial_\lambda g_{\delta\beta} + g^{\lambda\delta}\partial_\beta\partial_\lambda g_{\delta\alpha} - g^{\lambda\delta}\partial_\alpha\partial_\beta g_{\lambda\delta} \\
&\quad + \partial_\alpha g^{\lambda\delta}\partial_\delta g_{\lambda\beta} + \partial_\beta g^{\lambda\delta}\partial_\delta g_{\lambda\alpha} - \frac{1}{2}\partial_\beta g^{\lambda\delta}\partial_\alpha g_{\lambda\delta} - \frac{1}{2}\partial_\alpha g^{\lambda\delta}\partial_\beta g_{\lambda\delta}.
\end{aligned}
\tag{4.6}
$$

We observe that the last term in (4.6) coincides with the last term in (4.5). By noting also that the second-order terms in $\partial_\alpha\Gamma_\beta + \partial_\beta\Gamma_\alpha$ are exactly three of the (four) second-order terms arising in the expression of $\partial_\lambda\Gamma^\lambda_{\alpha\beta} - \partial_\alpha\Gamma^\lambda_{\beta\lambda}$, we see that the term $\partial_\lambda\Gamma^\lambda_{\alpha\beta} - \partial_\alpha\Gamma^\lambda_{\beta\lambda}$ equals

$$
\begin{aligned}
&-\frac{1}{2} g^{\lambda\delta}\partial_\lambda\partial_\delta g_{\alpha\beta} + \frac{1}{2}\big(\partial_\alpha\Gamma_\beta + \partial_\beta\Gamma_\alpha\big) \\
&\quad - \frac{1}{2}\partial_\lambda g^{\lambda\delta}\partial_\delta g_{\alpha\beta} + \frac{1}{2}\partial_\lambda g^{\lambda\delta}\partial_\alpha g_{\beta\delta} + \frac{1}{2}\partial_\lambda g^{\lambda\delta}\partial_\beta g_{\alpha\delta} \\
&\quad - \frac{1}{2}\partial_\alpha g^{\lambda\delta}\partial_\delta g_{\lambda\beta} - \frac{1}{2}\partial_\beta g^{\lambda\delta}\partial_\delta g_{\lambda\alpha} - \frac{1}{4}\partial_\alpha g^{\lambda\delta}\partial_\beta g_{\lambda\delta} + \frac{1}{4}\partial_\beta g^{\lambda\delta}\partial_\alpha g_{\lambda\delta} \\
&= -\frac{1}{2}\partial_\lambda g^{\lambda\delta}\partial_\delta g_{\alpha\beta} + \frac{1}{2}\big(\partial_\alpha\Gamma_\beta + \partial_\beta\Gamma_\alpha\big) \\
&\quad + \frac{1}{2} g^{\lambda\lambda'} g^{\delta\delta'}\partial_\lambda g_{\lambda'\delta'}\partial_\delta g_{\alpha\beta} - \frac{1}{2} g^{\lambda\lambda'} g^{\delta\delta'}\partial_\lambda g_{\lambda'\delta'}\partial_\alpha g_{\beta\delta} \\
&\quad - \frac{1}{2} g^{\lambda\lambda'} g^{\delta\delta'}\partial_\lambda g_{\lambda'\delta'}\partial_\beta g_{\alpha\delta} + \underline{\frac{1}{4} g^{\lambda\lambda'} g^{\delta\delta'}\partial_\alpha g_{\lambda'\delta'}\partial_\beta g_{\lambda\delta}} \\
&\quad + \frac{1}{2} g^{\lambda\lambda'} g^{\delta\delta'}\partial_\alpha g_{\lambda'\delta'}\partial_\delta g_{\lambda\beta} + \frac{1}{2} g^{\lambda\lambda'} g^{\delta\delta'}\partial_\beta g_{\lambda'\delta'}\partial_\delta g_{\lambda\alpha} - \underline{\frac{1}{4} g^{\lambda\lambda'} g^{\delta\delta'}\partial_\beta g_{\lambda'\delta'}\partial_\alpha g_{\lambda\delta}},
\end{aligned}
$$

where we have used the identity $\partial_\alpha g^{\lambda\delta} = -g^{\lambda\lambda'} g^{\delta\delta'} \partial_\alpha g_{\lambda'\delta'}$. Note that the two underlined terms above cancel each other. So, the quadratic terms in $\partial_\lambda \Gamma^\lambda_{\alpha\beta} - \partial_\alpha \Gamma^\lambda_{\beta\lambda}$ are

$$\frac{1}{2} g^{\lambda\lambda'} g^{\delta\delta'} \partial_\lambda g_{\lambda'\delta'} \partial_\delta g_{\alpha\beta}, \quad -\frac{1}{2} g^{\lambda\lambda'} g^{\delta\delta'} \partial_\lambda g_{\lambda'\delta'} \partial_\alpha g_{\beta\delta}, \quad -\frac{1}{2} g^{\lambda\lambda'} g^{\delta\delta'} \partial_\lambda g_{\lambda'\delta'} \partial_\beta g_{\alpha\delta},$$
$$\frac{1}{2} g^{\lambda\lambda'} g^{\delta\delta'} \partial_\alpha g_{\lambda'\delta'} \partial_\delta g_{\lambda\beta}, \quad \frac{1}{2} g^{\lambda\lambda'} g^{\delta\delta'} \partial_\beta g_{\lambda'\delta'} \partial_\delta g_{\lambda\alpha}.$$

Next, let us return to the expression of the Ricci curvature and consider

$$\begin{aligned}
\Gamma^\lambda_{\alpha\beta} \Gamma^\delta_{\lambda\delta} &= \frac{1}{4} g^{\lambda\lambda'} g^{\delta\delta'} \big(\partial_\lambda g_{\delta\delta'} \partial_\alpha g_{\beta\lambda'} + \partial_\beta g_{\alpha\lambda'} \partial_\lambda g_{\delta\delta'} - \partial_{\lambda'} g_{\alpha\beta} \partial_\lambda g_{\delta\delta'} \big), \\
\Gamma^\lambda_{\alpha\delta} \Gamma^\delta_{\beta\lambda} &= \frac{1}{4} g^{\lambda\lambda'} g^{\delta\delta'} \big(\partial_\alpha g_{\delta\lambda'} \partial_\beta g_{\lambda\delta'} + \partial_\alpha g_{\delta\lambda'} \partial_\lambda g_{\beta\delta'} - \partial_\alpha g_{\delta\lambda'} \partial_{\delta'} g_{\beta\lambda} \\
&\quad + \partial_\delta g_{\alpha\lambda'} \partial_\beta g_{\lambda\delta'} + \partial_\delta g_{\alpha\lambda'} \partial_\lambda g_{\beta\delta'} - \partial_\delta g_{\alpha\lambda'} \partial_{\delta'} g_{\beta\lambda} \\
&\quad - \partial_{\lambda'} g_{\alpha\delta} \partial_\beta g_{\lambda\delta'} - \partial_{\lambda'} g_{\alpha\delta} \partial_\lambda g_{\beta\delta'} + \partial_{\lambda'} g_{\alpha\delta} \partial_{\delta'} g_{\beta\lambda} \big)
\end{aligned}$$

and deduce that

$$\begin{aligned}
&\Gamma^\lambda_{\alpha\beta} \Gamma^\delta_{\lambda\delta} - \Gamma^\lambda_{\alpha\delta} \Gamma^\delta_{\beta\lambda} \\
&= -\frac{1}{4} g^{\lambda\lambda'} g^{\delta\delta'} \partial_{\lambda'} g_{\alpha\beta} \partial_\lambda g_{\delta\delta'} + \frac{1}{4} g^{\lambda\lambda'} g^{\delta\delta'} \partial_\delta g_{\alpha\lambda'} \partial_{\delta'} g_{\beta\lambda} + \frac{1}{4} g^{\lambda\lambda'} \partial_{\lambda'} g_{\alpha\delta} \partial_\lambda g_{\beta\delta'} \\
&\quad - \frac{1}{4} g^{\lambda\lambda'} g^{\delta\delta'} \partial_\alpha g_{\delta\lambda'} \partial_\beta g_{\lambda\delta'} \\
&\quad + \underline{\frac{1}{4} g^{\lambda\lambda'} g^{\delta\delta'} \partial_\lambda g_{\delta\delta'} \partial_\alpha g_{\beta\lambda'}} + \underline{\frac{1}{4} g^{\lambda\lambda'} g^{\delta\delta'} \partial_\lambda g_{\delta\delta'} \partial_\beta g_{\alpha\lambda'}} - \frac{1}{2} g^{\lambda\lambda'} g^{\delta\delta'} \partial_\delta g_{\alpha\lambda'} \partial_\lambda g_{\beta\delta'}.
\end{aligned} \tag{4.7}$$

Observe that the first three terms are null terms, while the fourth term is a quasi-null term. The two underlined terms are going to cancel out with the two underlined terms in (4.10), derived below. Hence, there remains only the last term to be treated.

In other words, we need to consider the following six terms:

$$\begin{aligned}
&\frac{1}{2} g^{\lambda\lambda'} g^{\delta\delta'} \partial_\lambda g_{\lambda'\delta'} \partial_\delta g_{\alpha\beta}, \quad -\frac{1}{2} g^{\lambda\lambda'} g^{\delta\delta'} \partial_\lambda g_{\lambda'\delta'} \partial_\alpha g_{\beta\delta}, \quad -\frac{1}{2} g^{\lambda\lambda'} g^{\delta\delta'} \partial_\lambda g_{\lambda'\delta'} \partial_\beta g_{\alpha\delta}, \\
&\frac{1}{2} g^{\lambda\lambda'} g^{\delta\delta'} \partial_\alpha g_{\lambda'\delta'} \partial_\delta g_{\lambda\beta}, \quad \frac{1}{2} g^{\lambda\lambda'} g^{\delta\delta'} \partial_\beta g_{\lambda'\delta'} \partial_\delta g_{\lambda\alpha}, \quad -\frac{1}{2} g^{\lambda\lambda'} g^{\delta\delta'} \partial_\delta g_{\alpha\lambda'} \partial_\lambda g_{\beta\delta'}.
\end{aligned} \tag{4.8}$$

In view of the identities

$$g^{\alpha\beta} \partial_\alpha g_{\beta\delta} - \frac{1}{2} g^{\alpha\beta} \partial_\delta g_{\alpha\beta} = \Gamma_\delta, \qquad g_{\beta\delta} \partial_\alpha g^{\alpha\beta} - \frac{1}{2} g_{\alpha\beta} \partial_\delta g^{\alpha\beta} = \Gamma_\delta, \tag{4.9}$$

the first three terms in (4.8) can be decomposed as follows:

$$
\begin{aligned}
\frac{1}{2}g^{\lambda\lambda'}g^{\delta\delta'}\partial_\lambda g_{\lambda'\delta'}\partial_\delta g_{\alpha\beta} &= \frac{1}{2}g^{\delta\delta'}\partial_\delta g_{\alpha\beta}\Gamma_{\delta'} + \frac{1}{4}g^{\lambda\lambda'}g^{\delta\delta'}\partial_\delta g_{\alpha\beta}\partial_{\delta'}g_{\lambda\lambda'} \\
-\frac{1}{2}g^{\lambda\lambda'}g^{\delta\delta'}\partial_\lambda g_{\lambda'\delta'}\partial_\alpha g_{\beta\delta} &= -\frac{1}{2}g^{\delta\delta'}\partial_\alpha g_{\beta\delta}\Gamma_{\delta'} - \underline{\frac{1}{4}g^{\lambda\lambda'}g^{\delta\delta'}\partial_{\delta'}g_{\lambda\lambda'}\partial_\alpha g_{\beta\delta}} \\
-\frac{1}{2}g^{\lambda\lambda'}g^{\delta\delta'}\partial_\lambda g_{\lambda'\delta'}\partial_\beta g_{\alpha\delta} &= -\frac{1}{2}g^{\delta\delta'}\partial_\beta g_{\alpha\delta}\Gamma_{\delta'} - \underline{\frac{1}{4}g^{\lambda\lambda'}g^{\delta\delta'}\partial_{\delta'}g_{\lambda\lambda'}\partial_\beta g_{\alpha\delta}}.
\end{aligned}
\tag{4.10}
$$

The last term in the first line is one of the quasi-null term stated in the proposition. As mentioned earlier, the two underlined terms cancel out with the two underlined terms in (4.7). The fourth term in (4.8) is treated as follows:

$$
\begin{aligned}
&\frac{1}{2}g^{\lambda\lambda'}g^{\delta\delta'}\partial_\alpha g_{\lambda'\delta'}\partial_\delta g_{\lambda\beta} \\
&= \frac{1}{2}g^{\lambda\lambda'}g^{\delta\delta'}\big(\partial_\alpha g_{\lambda'\delta'}\partial_\delta g_{\lambda\beta} - \partial_\alpha g_{\lambda\beta}\partial_\delta g_{\lambda'\delta'}\big) + \frac{1}{2}g^{\lambda\lambda'}g^{\delta\delta'}\partial_\alpha g_{\lambda\beta}\partial_\delta g_{\lambda'\delta'} \\
&= \frac{1}{2}g^{\lambda\lambda'}g^{\delta\delta'}\big(\partial_\alpha g_{\lambda'\delta'}\partial_\delta g_{\lambda\beta} - \partial_\alpha g_{\lambda\beta}\partial_\delta g_{\lambda'\delta'}\big) + \frac{1}{2}g^{\lambda\lambda'}\partial_\alpha g_{\lambda\beta}\Gamma_{\lambda'} + \frac{1}{4}g^{\lambda\lambda'}g^{\delta\delta'}\partial_\alpha g_{\lambda\beta}\partial_{\lambda'}g_{\delta\delta'} \\
&= \frac{1}{2}g^{\lambda\lambda'}g^{\delta\delta'}\big(\partial_\alpha g_{\lambda'\delta'}\partial_\delta g_{\lambda\beta} - \partial_\alpha g_{\lambda\beta}\partial_\delta g_{\lambda'\delta'}\big) + \frac{1}{4}g^{\lambda\lambda'}g^{\delta\delta'}\big(\partial_\alpha g_{\lambda\beta}\partial_{\lambda'}g_{\delta\delta'} - \partial_\alpha g_{\delta\delta'}\partial_{\lambda'}g_{\lambda\beta}\big) \\
&\quad + \frac{1}{2}g^{\lambda\lambda'}\partial_\alpha g_{\lambda\beta}\Gamma_{\lambda'} + \frac{1}{4}g^{\lambda\lambda'}g^{\delta\delta'}\partial_\alpha g_{\delta\delta'}\partial_{\lambda'}g_{\lambda\beta} \\
&= \frac{1}{2}g^{\lambda\lambda'}g^{\delta\delta'}\big(\partial_\alpha g_{\lambda'\delta'}\partial_\delta g_{\lambda\beta} - \partial_\alpha g_{\lambda\beta}\partial_\delta g_{\lambda'\delta'}\big) + \frac{1}{4}g^{\lambda\lambda'}g^{\delta\delta'}\big(\partial_\alpha g_{\lambda\beta}\partial_{\lambda'}g_{\delta\delta'} - \partial_\alpha g_{\delta\delta'}\partial_{\lambda'}g_{\lambda\beta}\big) \\
&\quad + \frac{1}{2}g^{\lambda\lambda'}\partial_\alpha g_{\lambda\beta}\Gamma_{\lambda'} + \frac{1}{4}g^{\delta\delta'}\partial_\alpha g_{\delta\delta'}\Gamma_\beta + \frac{1}{8}g^{\delta\delta'}g^{\lambda\lambda'}\partial_\alpha g_{\delta\delta'}\partial_\beta g_{\lambda\lambda'},
\end{aligned}
$$

while, for the fifth term, we have

$$
\begin{aligned}
&\frac{1}{2}g^{\lambda\lambda'}g^{\delta\delta'}\partial_\beta g_{\lambda'\delta'}\partial_\delta g_{\lambda\alpha} \\
&= \frac{1}{2}g^{\lambda\lambda'}g^{\delta\delta'}\big(\partial_\beta g_{\lambda'\delta'}\partial_\delta g_{\lambda\alpha} - \partial_\beta g_{\lambda\alpha}\partial_\delta g_{\lambda'\delta'}\big) \\
&+ \frac{1}{4}g^{\lambda\lambda'}g^{\delta\delta'}\big(\partial_\beta g_{\lambda\alpha}\partial_{\lambda'}g_{\delta\delta'} - \partial_\beta g_{\delta\delta'}\partial_{\lambda'}g_{\lambda\alpha}\big) \\
&\quad + \frac{1}{2}g^{\lambda\lambda'}\partial_\beta g_{\lambda\alpha}\Gamma_{\lambda'} + \frac{1}{4}g^{\delta\delta'}\partial_\beta g_{\delta\delta'}\Gamma_\alpha + \frac{1}{8}g^{\delta\delta'}g^{\lambda\lambda'}\partial_\beta g_{\delta\delta'}\partial_\alpha g_{\lambda\lambda'}.
\end{aligned}
$$

For the last term in (4.8), we perform the following calculation:

$$
\begin{aligned}
&-\frac{1}{2}g^{\lambda\lambda'}g^{\delta\delta'}\partial_\delta g_{\alpha\lambda'}\partial_\lambda g_{\beta\delta'}\\
&= -\frac{1}{2}g^{\lambda\lambda'}g^{\delta\delta'}\left(\partial_\delta g_{\alpha\lambda'}\partial_\lambda g_{\beta\delta'} - \partial_\delta g_{\beta\delta'}\partial_\lambda g_{\alpha\lambda'}\right) - \frac{1}{2}g^{\lambda\lambda'}g^{\delta\delta'}\partial_\delta g_{\beta\delta'}\partial_\lambda g_{\alpha\lambda'}\\
&= -\frac{1}{2}g^{\lambda\lambda'}g^{\delta\delta'}\left(\partial_\delta g_{\alpha\lambda'}\partial_\lambda g_{\beta\delta'} - \partial_\delta g_{\beta\delta'}\partial_\lambda g_{\alpha\lambda'}\right)\\
&\quad - \frac{1}{2}g^{\lambda\lambda'}\partial_\lambda g_{\alpha\lambda'}\Gamma_\beta - \frac{1}{4}g^{\lambda\lambda'}g^{\delta\delta'}\partial_\beta g_{\delta\delta'}\partial_\lambda g_{\alpha\lambda'}\\
&= -\frac{1}{2}g^{\lambda\lambda'}g^{\delta\delta'}\left(\partial_\delta g_{\alpha\lambda'}\partial_\lambda g_{\beta\delta'} - \partial_\delta g_{\beta\delta'}\partial_\lambda g_{\alpha\lambda'}\right)\\
&\quad - \frac{1}{2}g^{\lambda\lambda'}\partial_\lambda g_{\alpha\lambda'}\Gamma_\beta - \frac{1}{4}g^{\delta\delta'}\partial_\beta g_{\delta\delta'}\Gamma_\alpha\\
&\quad - \frac{1}{8}g^{\lambda\lambda'}g^{\delta\delta'}\partial_\alpha g_{\lambda\lambda'}\partial_\beta g_{\delta\delta'}\\
&= -\frac{1}{2}g^{\lambda\lambda'}g^{\delta\delta'}\left(\partial_\delta g_{\alpha\lambda'}\partial_\lambda g_{\beta\delta'} - \partial_\delta g_{\beta\delta'}\partial_\lambda g_{\alpha\lambda'}\right) - \frac{1}{2}\Gamma_\alpha\Gamma_\beta\\
&\quad - \frac{1}{4}g^{\delta\delta'}\partial_\alpha g_{\delta\delta'}\Gamma_\beta - \frac{1}{4}g^{\delta\delta'}\partial_\beta g_{\delta\delta'}\Gamma_\alpha\\
&\quad - \frac{1}{8}g^{\lambda\lambda'}g^{\delta\delta'}\partial_\alpha g_{\lambda\lambda'}\partial_\beta g_{\delta\delta'}.
\end{aligned}
$$

In conclusion, the quadratic terms in $R_{\alpha\beta}$ read

$$
\begin{aligned}
&\frac{1}{2}g^{\lambda\lambda'}g^{\delta\delta'}\partial_\delta g_{\alpha\lambda'}\partial_{\delta'}g_{\beta\lambda}\\
&\quad - \frac{1}{2}g^{\lambda\lambda'}g^{\delta\delta'}\left(\partial_\delta g_{\alpha\lambda'}\partial_\lambda g_{\beta\delta'} - \partial_\delta g_{\beta\delta'}\partial_\lambda g_{\alpha\lambda'}\right)\\
&\quad + \frac{1}{2}g^{\lambda\lambda'}g^{\delta\delta'}\left(\partial_\alpha g_{\lambda'\delta'}\partial_\delta g_{\lambda\beta} - \partial_\alpha g_{\lambda\beta}\partial_\delta g_{\lambda'\delta'}\right)\\
&\quad + \frac{1}{4}g^{\lambda\lambda'}g^{\delta\delta'}\left(\partial_\alpha g_{\lambda\beta}\partial_{\lambda'}g_{\delta\delta'} - \partial_\alpha g_{\delta\delta'}\partial_{\lambda'}g_{\lambda\beta}\right)\\
&\quad + \frac{1}{2}g^{\lambda\lambda'}g^{\delta\delta'}\left(\partial_\beta g_{\lambda'\delta'}\partial_\delta g_{\lambda\alpha} - \partial_\beta g_{\lambda\alpha}\partial_\delta g_{\lambda'\delta'}\right)\\
&\quad + \frac{1}{4}g^{\lambda\lambda'}g^{\delta\delta'}\left(\partial_\beta g_{\lambda\alpha}\partial_{\lambda'}g_{\delta\delta'} - \partial_\beta g_{\delta\delta'}\partial_{\lambda'}g_{\lambda\alpha}\right)\\
&\quad - \frac{1}{4}g^{\lambda\lambda'}g^{\delta\delta'}\partial_\alpha g_{\delta\lambda'}\partial_\beta g_{\lambda\delta'} + \frac{1}{8}g^{\delta\delta'}g^{\lambda\lambda'}\partial_\beta g_{\delta\delta'}\partial_\alpha g_{\lambda\lambda'}\\
&\quad + \frac{1}{2}g^{\delta\delta'}\partial_\delta g_{\alpha\beta}\Gamma_{\delta'} - \frac{1}{2}\Gamma_\alpha\Gamma_\beta.
\end{aligned}
$$

Finally, collecting all the terms above and observing that several cancellations take place, we arrive at the desired identity. □

4.2 Analysis of the support

We provide here a proof of Proposition 2.1.

Step I. We recall the structure of $F_{\alpha\beta}$ presented in Lemma 4.1. We observe that both $P_{\alpha\beta}$ and $Q_{\alpha\beta}$ are linear combinations of the multi-linear terms which are product of a quadratic term in $g^{\alpha\beta}$ and a quadratic term in $\partial g_{\alpha\beta}$. For convenience, we write $F_{\alpha\beta} = F_{\alpha\beta}(g, g; \partial g, \partial g)$ and

$$p_{\alpha\beta}(t,x) := \big(g_{S\alpha\beta} - m_{\alpha\beta}\big)(t,x)\xi(t-r) + m_{\alpha\beta},$$

where ξ is a smooth function defined on $\mathbb{R}$, with $\xi(r) = 1$ for $r \leq 1$, while $\xi(r) = 0$ for $r \geq 3/2$. Hence, for $r \geq t-1$, $p_{\alpha\beta}$ coincides with the Schwarzschild metric while $r \leq t - 3/2$, $p_{\alpha\beta}$ coincides with the Minkowski metric. We also set

$$q_{\alpha\beta} := g_{\alpha\beta} - p_{\alpha\beta}. \tag{4.11}$$

So the desired result is equivalent to the following statement: *If $(g_{\alpha\beta}, \phi)$ is a solution of (4.4) associated with a compact Schwarzschild perturbation, then the tensor $q_{\alpha\beta}$ above is supported in $\mathcal{K}$.*

To establish this result, we write down the equation satisfied by $q_{\alpha\beta}$ and introduce

$$\begin{aligned}
(p^{\alpha\beta}) :=& (p_{\alpha\beta})^{-1},\\
q^{\alpha\beta} :=& g^{\alpha\beta} - p^{\alpha\beta} = (p_{\alpha'\beta'} - g_{\alpha'\beta'})p^{\alpha'\beta}g_{\alpha\beta'} = q_{\alpha'\beta'}p^{\alpha'\beta}g^{\alpha\beta'}.
\end{aligned}$$

We observe that for $r \geq t-1$, when $q_{\alpha\beta}(t,x) = 0$, then $q^{\alpha\beta}(t,x) = 0$. In view of

$$\widetilde{\Box}_g g_{\alpha\beta} = F_{\alpha\beta}(g, g, \partial g, \partial g) - 16\pi\partial_\alpha\phi\partial_\beta\phi - 8\pi c^2\phi^2 g_{\alpha\beta},$$

we have

$$\widetilde{\Box}_{p+q}(p_{\alpha\beta}+q_{\alpha\beta}) = F_{\alpha\beta}\big(p+q, p+q, \partial(p+q), \partial(p+q)\big) - 16\pi\partial_\alpha\phi\partial_\beta\phi - 8\pi c^2\phi^2 g_{\alpha\beta}.$$

By multi-linearity, the above equation leads us to

$$\begin{aligned}
\widetilde{\Box}_p q_{\alpha\beta} =& -\widetilde{\Box}_p p_{\alpha\beta} + F_{\alpha\beta}\big(p, p, \partial p, \partial p\big)\\
&+ F_{\alpha\beta}\big(p, p, \partial p, \partial q\big) + F_{\alpha\beta}\big(p, p, \partial q, \partial(p+q)\big)\\
&+ F_{\alpha\beta}\big(p, q, \partial(p+q), \partial(p+q)\big) + F_{\alpha\beta}\big(q, p+q, \partial(p+q), \partial(p+q)\big)\\
&- q^{\mu\nu}\partial_\mu\partial_\nu\big(p_{\alpha\beta} + q_{\alpha\beta}\big) - 16\pi\partial_\alpha\phi\partial_\beta\phi - 8\pi c^2\phi^2 g_{\alpha\beta}.
\end{aligned} \tag{4.12}$$

Observe that for $r \geq t-1$, $p_{\alpha\beta} = \big(g_{S\alpha\beta} - m_{\alpha\beta}\big)\xi(t-r) + m_{\alpha\beta}$ coincides with the Schwarzschild metric, which is a solution to the Einstein equation (in the wave gauge), so for $r \geq t-1$ we have $\widetilde{\Box}_p p_{\alpha\beta} = F_{\alpha\beta}(p,p,\partial p,\partial p)$. Setting $E_{\alpha\beta} = -\widetilde{\Box}_p p_{\alpha\beta} + F_{\alpha\beta}(p,p,\partial p,\partial p)$, we have obtained $E_{\alpha\beta}=0$ for $r \geq t-1$.

Then we also observe that the third to the sixth terms are multi-linear terms, each of them contain q or ∂q as a factor. Furthermore, we observe that the seventh term is written as

$$-q^{\mu\nu}\partial_\mu\partial_\nu(p_{\alpha\beta}+q_{\alpha\beta}) = -q_{\mu'\nu'}p^{\mu'\nu}g^{\mu\nu'}\partial_\mu\partial_\nu(p_{\alpha\beta}+q_{\alpha\beta}).$$

So, the third to the seventh terms can be written in the form

$$\partial q \cdot G_1(p,\partial p,q,\partial q) + q \cdot G_2(p,\partial p,\partial\partial p,q,\partial q),$$

where G_i are (sufficiently regular) multi-linear forms.

For the equation of ϕ, we have the decomposition

$$\widetilde{\Box}_g\phi = \Box_p\phi + \widetilde{\Box}_q\phi = \widetilde{\Box}_p\phi + q_{\mu'\nu'}p^{\mu'\nu}g^{\mu\nu'}\partial_\mu\partial_\nu\phi.$$

We conclude that the metric $q_{\alpha\beta}$ satisfies

$$\begin{aligned}
\widetilde{\Box}_p q_{\alpha\beta} &= E_{\alpha\beta} + \partial q \cdot G_1(p,\partial p,q,\partial q) + q \cdot G_2(p,\partial p,\partial\partial p,q,\partial q) \\
&\quad - 16\pi\partial_\alpha\phi\partial_\beta\phi - 8\pi c^2\phi^2 g_{\alpha\beta}, \\
\widetilde{\Box}_p\phi - c^2\phi &= -q_{\mu'\nu'}p^{\mu'\nu}g^{\mu\nu'}\partial_\mu\partial_\nu\phi.
\end{aligned} \tag{4.13}$$

Furthermore, observe that since (g,ϕ) describes a compact Schwarzschild perturbation, the restriction of both $q_{\alpha\beta}$ and ϕ on the hyperplane $\{t=2\}$ are compactly supported in the unit ball $\{r \leq 1\}$. Thus, $(q_{\alpha\beta},\phi)$ is a regular solution to the linear wave system (4.13) with initial data

$$q_{\alpha\beta}(2,x), \quad \phi(2,x) \qquad \text{supported in the ball } \{r \leq 1\}.$$

We want to prove that $(q_{\alpha\beta})$ and ϕ vanish outside $\mathcal{K}$. This leads us to the analysis on the domain of determinacy associated with the metric $p^{\alpha\beta}$, which is determined by the characteristics the operator $\widetilde{\Box}_p$.

Step II. Characteristics of $\widetilde{\Box}_p$. We now analyze the domain of determinacy of a spacetime point $(t,x) \notin \mathcal{K}$. We will prove that all characteristics passing this point do not intersect the domain $\mathcal{K} \cap \{t \geq 2\}$. Once this is proved, we apply the standard argument on domain of determinacy (also observe that $E_{\alpha\beta}(t,x)$ vanishes outside $\mathcal{K}$), we conclude that $q_{\alpha\beta}$ and ϕ vanish outside $\mathcal{K}$.

To do so, we will prove that the boundary of $\mathcal{K}$ is strictly spacelike with respect to the metric $p^{\alpha\beta}$. We observe that any vector v tangent to

$\{r = t-1\}$ at point (t, x) satisfies $v^0 = \frac{1}{r}\sum_a x^a v^a = \omega_a v^a$. So, in view of (2.6), we have for all $|v| > 0$

$$
\begin{aligned}
(v, v)_p(t, x) &= (v, v)_{g_S} = v^0 v^0 g_{00} + v^a v^b g_{ab} \\
&= -\frac{r - m_S}{r + m_S}\omega_a v^a \omega_b v^b + \omega_a v^a \omega_b v^b \left(\frac{r + m_S}{r - m_S} - \frac{(r + m_S)^2}{r^2}\right) + \sum_a |v^a|^2 \\
&= -\left(\frac{r - m_S}{r + m_S} - \frac{r + m_S}{r - m_S} + \frac{(r + m_S)^2}{r^2}\right)\omega_a v^a \omega_b v^b + \sum_a |v^a|^2 \\
&\geq \left(1 - \left(\frac{r + m_S}{r - m_S} - \frac{r - m_S}{r + m_S} + \frac{r^2}{(r + m_S)^2}\right)\omega_a v^a \omega_b v^b\right)\sum_a |v^a|^2 \\
&= \frac{3r^2 m_S + 4r m_S^2 + m_S^3}{(r + m_S)^2 (r - m_S)}\sum_a |v^a|^2 > 0,
\end{aligned}
$$

where we have used $|\omega_a v^a| \leq |v| = \left(\sum_a |v^a|^2\right)^{1/2}$.

A characteristic curve is a null curve, so a characteristic passing through (t, x) with $r \geq t-1$ cannot intersect the boundary $\{r = t-1\}$ in the past direction (since (t, x) is already in the past of $\{r = t-1\}$). Hence, a characteristic passing through (t, x) never intersects the region $\mathcal{K}$ in the past direction, which leads to the conclusion that the domain of determinacy of (t, x) does not intersect $\mathcal{K}$ and, therefore, does not intersect $\{t = 2, r \leq t-1\}$. We conclude that $q_{\alpha\beta}(t, x) = \phi(t, x) = 0$.

4.3 A classification of nonlinearities in the Einstein-massive field system

First, we introduce a class of functions of particular interest.

Definition 4.1. A smooth and homogeneous function (defined in $\{r < t\}$) of degree α is, by definition, a smooth function Φ defined in $\{r < t\}$ at least and satisfying

- $\Phi(\lambda t, \lambda x) = \lambda^\alpha \Phi(t, x)$, for a fixed $\alpha \in \mathbb{R}$ and for all $\lambda > 0$,
- $\sup_{|x| \leq 1} |\partial^I \Phi(1, x)| < +\infty$ (for large enough $|I|$).

For instance, constant functions are smooth and homogeneous functions of degree 0. We also observe that the elements of the transition matrix Φ_α^β are smooth and homogeneous of degree 0.

Lemma 4.2. *If Φ is a smooth and homogeneous function defined in $\{r \leq t\}$ of degree α, then there exists a constant C determined by Φ and N such that*

$$|\partial^I L^J \Phi(t, x)| \leq C t^{\alpha - |I|}.$$

Furthermore, if Φ and Ψ are smooth and homogenous functions of degree α and β, respectively, then the product $\Phi\Psi$ is smooth and homogeneous of degree $(\alpha+\beta)$.

Proof. Observe that if Φ is homogeneous of degree α, then $\Phi(\lambda t, \lambda x) = \lambda^\alpha \Phi(t,x)$. We differentiate the above equation with respect to x^a: $\lambda \partial_a \Phi(\lambda t, \lambda x) = \lambda^\alpha \partial_a \Phi(t,x)$, which leads to $\partial_a \Phi(\lambda t, \lambda x) = \lambda^{\alpha-1}\partial_a \Phi(t,x)$. In the same way, we obtain $\partial_t \Phi(\lambda t, \lambda x) = \lambda^{\alpha-1}\partial_t \Phi(t,x)$. For L_a, we have

$$\begin{aligned} L_a \Phi(\lambda t, \lambda x) &= (\lambda x^a)\partial_t \Phi(\lambda t, \lambda x) + (\lambda t)\partial_a \Phi(\lambda t, \lambda x) \\ &= (\lambda x^a)\lambda^{\alpha-1}\partial_t \Phi(t,x) + (\lambda t)\lambda^{\alpha-1}\partial_a \Phi(t,x) \lambda^\alpha L_a \Phi(t,x). \end{aligned}$$

We conclude that, after differentiation by ∂_α, the degree of a homogeneous function will be reduced by one while when derived by L_a the degree does not change. By induction, we get the desired estimate. Finally, we observe that the relation between homogeneity and multiplication is trivial. □

In the following, the nonlinear terms such as $F_{\alpha\beta}$ and $[\partial^I L^J, h^{\mu\nu}\partial_\mu\partial_\nu]h_{\alpha\beta}$ are expressed as linear combinations of some basic nonlinear terms (presented below) with smooth and homogeneous coefficients of non-positive degrees. We provide first a general classification of such nonlinear terms:

- $QS_h(p,k)$ refers to at most *p-order quadratic semi-linear terms in $h_{\alpha\beta}$*. They are linear combinations of the following terms with smooth and homogeneous coefficients of degree ≤ 0:

$$\partial^I L^J \big(\partial_\mu h_{\alpha\beta} \partial_\nu h_{\alpha'\beta'}\big)$$

  with $|I|+|J| \leq p$, $|J| \leq k$.
- $QS_\phi(p,k)$ refers to *p-order quadratic semi-linear terms in ϕ*. They are linear combinations of the following terms with smooth and homogeneous coefficients of degree ≤ 0:

$$\partial^I L^J \big(\partial_\mu \phi \partial_\nu \phi\big), \quad \partial^I L^J (\phi^2 g_{\mu\nu})$$

  with $|I|+|J| \leq p$, $|J| \leq k$.
- $QQ_{hh}(p,k)$ refers to *p-order quadratic quasi-linear terms in h*, which arise from the expression $[\partial^I L^J, h^{\mu\nu}\partial_\mu\partial_\nu]h_{\alpha\beta}$. They are linear combinations of the following terms with smooth and homogeneous coefficients of degree ≤ 0:

$$\partial^{I_1} L^{J_1} h_{\alpha'\beta'} \partial^{I_2} L^{J_2} \partial_\mu \partial_\nu h_{\alpha\beta}, \qquad h_{\alpha'\beta'} \partial_\mu\partial_\nu \partial^I L^{J'} h_{\alpha\beta}$$

  with $|I_1|+|I_2| \leq p-k$, $|J_1|+|J_2| \leq k$ and $|I_2|+|J_2| \leq p-1$ and $|J'| < |J|$.

- $QQ_{h\phi}(p,k)$ refers to *p-order quadratic quasi-linear terms in h and ϕ.* These terms come from the commutator $[\partial^I L^J, h^{\mu\nu}\partial_\mu\partial_\nu]\phi$. They are linear combination of the following terms with smooth and homogeneous coefficients of degree ≤ 0:

$$\partial^{I_1}L^{J_1}h_{\alpha'\beta'}\partial^{I_2}L^{J_2}\partial_\mu\partial_\nu\phi, \qquad h_{\alpha'\beta'}\partial_\mu\partial_\nu\partial^I L^{J'}\phi$$

  with $|I_1|+|I_2| \leq p-k$, $|J_1|+|J_2| \leq k$ and $|I_2|+|J_2| \leq p-1$, $|J'| < |J|$.

Next, we provide a list of "good" nonlinear terms:

- $Cub(p,k)$ refers to *higher-order terms of at least cubic order, except* the cubic term $h_{\alpha\beta}h_{\gamma\delta}h_{\mu\nu}$ which does not appear in our system. This class covers all cubic terms of interest, in view of the structure of the system under consideration. Moreover, these terms are "negligible" as far as the analysis of global existence is concerned.
- $GQS_h(p,k)$ refers to *"good" quadratic semi-linear terms in ∂h*, that are linear combinations of the following terms with smooth and homogeneous coefficients of degree ≤ 0:

$$\partial^I L^J\big(\underline{\partial}_a h_{\alpha\beta}\underline{\partial}_\gamma h_{\alpha'\beta'}\big), \quad (s/t)^2\partial^I L^J\big(\partial_t h_{\alpha\beta}\partial_t h_{\alpha'\beta'}\big)$$

  with $|I|+|J| \leq p$ and $|J| \leq k$.
- $GQQ_{hh}(p,k)$ refers to *"good" quadratic quasi-linear terms*, that are linear combinations of the following terms with smooth and homogeneous coefficients of degree ≤ 0:

$$\partial^{I_1}L^{J_1}h_{\alpha'\beta'}\partial^{I_2}L^{J_2}\underline{\partial}_a\underline{\partial}_\mu h_{\alpha\beta}, \qquad \partial^{I_1}L^{J_1}h_{\alpha'\beta'}\partial^{I_2}L^{J_2}\underline{\partial}_\mu\underline{\partial}_b h_{\alpha\beta},$$
$$h_{\alpha'\beta'}\partial^I L^{J'}\underline{\partial}_a\underline{\partial}_\mu h_{\alpha\beta}, \qquad h_{\alpha'\beta'}\partial^I L^{J'}\underline{\partial}_\mu\underline{\partial}_b h_{\alpha\beta}$$

  with $|I_1|+|I_2| \leq p-k$, $|J_1|+|J_2| \leq k$ and $|I_2|+|J_2| \leq p-1$, $|J'| < |J|$.
- $GQQ_{h\phi}(p,k)$ refers to *"good" quadratic quasi-linear terms*, that are linear combinations of the following terms with smooth and homogeneous coefficients of degree ≤ 0:

$$\partial^{I_1}L^{J_1}h_{\alpha'\beta'}\partial^{I_2}L^{J_2}\underline{\partial}_a\underline{\partial}_\mu\phi, \qquad \partial^{I_1}L^{J_1}h_{\alpha'\beta'}\partial^{I_2}L^{J_2}\underline{\partial}_\mu\underline{\partial}_b\phi,$$
$$h_{\alpha'\beta'}\partial^I L^{J'}\underline{\partial}_a\underline{\partial}_\mu\phi, \qquad h_{\alpha'\beta'}\partial^I L^{J'}\underline{\partial}_\mu\underline{\partial}_b\phi$$

  with $|I_1|+|I_2| \leq |I| = p-k$, $|J_1|+|J_2| \leq k$ and $|I_2|+|J_2| \leq p-1$, $|J'| < |J|$.

- $Com(p,k)$. These terms arise when we express a second-order derivative written in the canonical frame into the semi-hyperboloidal frame. Since the coefficients of the transition matrix Φ_α^β and Ψ_α^β are homogeneous of degree zero, and the commutators contain at least one derivative of these coefficients as a factor, these terms are linear combinations of the following terms with homogeneous coefficients of degree ≤ 0:

$$
\begin{array}{ll}
t^{-1}QS_h(p,k), & t^{-1}QS_\phi(p,k), \\
t^{-1}\partial^{I_1}L^{J_1}\partial_\mu h_{\alpha\beta}\partial^{I_2}L^{J_2}\partial_\nu\phi, & \\
t^{-1}\partial^{I_1}L^{J_1}h_{\mu\nu}\partial^{I_2}L^{J_2}\partial_\gamma h_{\mu'\nu'}, & t^{-2}\partial^{I_1}L^{J_1}h_{\mu\nu}\partial^{I_2}L^{J_2}\phi, \\
t^{-2}\partial^{I_1}L^{J_1}h_{\mu\nu}\partial^{I_2}L^{J_2}h_{\mu'\nu'}, &
\end{array}
$$

where $|I| \leq p-k, |J| \leq k$ and $|I_1| + |J_1| \leq p-1$, $|I_1| + |I_2| \leq p-k, |J_1| + |J_2| \leq k$.

With the above notation, we can decompose the commutator $[\partial^I L^J, h^{\mu\nu}\partial_\mu\partial_\nu]u$, as follows.

Lemma 4.3 (Decomposition of quasi-linear terms). *Let $|I| = p-k$ and $|J| = k$. Suppose $h^{\mu\nu}\partial_\mu\partial_\nu$ is a second-order operator with sufficiently regular coefficients. Then $[\partial^I L^J, h^{\mu\nu}\partial_\mu\partial_\nu]h_{\alpha\beta}$ is a linear combination of the following terms with smooth and homogeneous coefficients of degree 0:*

$$
\begin{array}{ll}
GQQ_{hh}(p,k), & t^{-1}\partial^{I_3}L^{J_3}h_{\mu\nu}\partial^{I_4}L^{J_4}\partial_\gamma h_{\mu'\nu'}, \\
\partial^{I_1}L^{J_1}\underline{h}^{00}\partial^{I_2}L^{J_2}\partial_t\partial_t h_{\alpha\beta}, & L^{J_1'}\underline{h}^{00}\partial^I L^{J_2'}\partial_t\partial_t h_{\alpha\beta}, \\
\underline{h}^{00}\partial_\gamma\partial_{\gamma'}\partial^I L^{J'}h_{\alpha\beta}, &
\end{array}
\tag{4.14}
$$

where $I_1 + I_2 = I, J_1 + J_2 = J$ with $|I_1| \geq 1$, $J_1' + J_2' = J$ with $|J_1'| \geq 1$ and $|J'| < |J|$, $|I_3| + |I_4| \leq |I|, |J_3| + |J_4| \leq |J|$.

Proof. We have

$$
\begin{aligned}
&[\partial^I L^J, h^{\mu\nu}\partial_\mu\partial_\nu]h_{\alpha\beta} = [\partial^I L^J, \underline{h}^{\mu\nu}\underline{\partial}_\mu\underline{\partial}_\nu]h_{\alpha\beta} + [\partial^I L^J, h^{\mu\nu}\partial_\mu\Psi_\nu^{\nu'}\underline{\partial}_{\nu'}]h_{\alpha\beta} \\
&= [\partial^I L^J, \underline{h}^{00}\partial_t\partial_t]h_{\alpha\beta} \\
&\quad + [\partial^I L^J, \underline{h}^{a0}\underline{\partial}_a\partial_t]h_{\alpha\beta} + [\partial^I L^J, \underline{h}^{0a}\partial_t\underline{\partial}_a]h_{\alpha\beta} + [\partial^I L^J, \underline{h}^{ab}\underline{\partial}_a\underline{\partial}_b]h_{\alpha\beta} \\
&\quad + [\partial^I L^J, h^{\mu\nu}\partial_\mu\Psi_\nu^{\nu'}\underline{\partial}_{\nu'}]h_{\alpha\beta}.
\end{aligned}
\tag{4.15}
$$

The second, third, and fourth terms are in class $GQQ_{hh}(p,k)$ ($\underline{h}^{\alpha\beta}$ being linear combinations of $h^{\alpha\beta}$ with smooth and homogeneous coefficients of degree zero) and, for the last term, we see that

$$\begin{aligned}
&[\partial^I L^J, h^{\mu\nu}\partial_\mu \Phi_\nu^{\nu'} \underline{\partial}_{\nu'}]h_{\alpha\beta} \\
&= \sum_{\substack{I_1+I_2+I_3=I \\ J_1+J_2+J_3=J \\ |I_3|+|J_3|<|I|+|J|}} \partial^{I_1} L^{J_1} h^{\mu\nu} \partial^{I_2} L^{J_2} \partial_\mu \Psi_\nu^{\nu'} \partial^{I_3} L^{J_3} \underline{\partial}_{\nu'} h_{\alpha\beta} \\
&\quad + h^{\mu\nu}\partial_\mu \Psi_\nu^{\nu'} [\partial^I L^J, \underline{\partial}_{\nu'}]h_{\alpha\beta}.
\end{aligned}$$

Then by the homogeneity of $\Psi_\nu^{\nu'}$, the above term can be expressed as $t^{-1}\partial^{I_3} L^{J_3} h_{\mu\nu} \partial^{I_4} L^{J_4} \partial_\gamma h_{\mu'\nu'}$.

Next, we treat the first term in the right-hand side of (4.15):

$$\begin{aligned}
&[\partial^I L^J, \underline{h}^{00}\partial_t\partial_t]h_{\alpha\beta} \\
&= \sum_{\substack{I_1+I_2=I \\ J_1+J_2=J, |I_1|\geq 1}} \partial^{I_1} L^{J_1} \underline{h}^{00} \partial^{I_2} L^{J_2} \partial_t\partial_t h_{\alpha\beta} + \sum_{\substack{J_1+J_2=J \\ |J_1|\geq 1}} L^{J_1} \underline{h}^{00} \partial^I L^{J_2} \partial_t\partial_t h_{\alpha\beta} \\
&\quad + \underline{h}^{00}[\partial^I L^J, \partial_t\partial_t]h_{\alpha\beta}.
\end{aligned}$$

We observe that $[\partial^I L^J, \partial_t\partial_t]h_{\alpha\beta}$ is a linear combination of the terms $\partial_{\alpha'}\partial_{\beta'}\partial^I L^{J'} h_{\alpha\beta}$ with $|J'| < |J|$. We apply the commutator identity (3.43):

$$\begin{aligned}
[\partial^I L^J, \partial_t\partial_t]h_{\alpha\beta} &= \partial^I [L^J, \partial_t\partial_t]h_{\alpha\beta} = \partial^I \left([L^J, \partial_t]\partial_t h_{\alpha\beta}\right) + \partial^I \partial_t \left([L^J, \partial_t]h_{\alpha\beta}\right) \\
&= \theta_{0J'}^{J\gamma} \partial_\gamma \partial_t L^{J'} h_{\alpha\beta} + \theta_{0J'}^{J\gamma} \theta_{0J''}^{J'\gamma'} \partial_{\gamma'} L^{J''} h_{\alpha\beta} + \theta_{0J'}^{J\gamma} \partial_t\partial_t L^{J'} h_{\alpha\beta},
\end{aligned}$$

where $|J''| < |J'| < |J|$. This completes the proof. □

A similar decomposition is available for the commutator

$$[\partial^I L^J, h^{\mu\nu}\partial_\mu\partial_\nu]\phi.$$

It is a linear combination of the following terms with smooth and homogeneous coefficients of degree ≤ 0:

$$\begin{aligned}
&GQQ_{h\phi}(p,k), \quad t^{-1}\partial^{I_1} L^{J_1} h_{\mu\nu} \partial^{I_2} L^{J_2} \partial_\gamma \phi, \\
&\partial^{I_1} L^{J_1} \underline{h}^{00} \partial^{I_2} L^{J_2} \partial_t\partial_t \phi, \quad L^{J_1'} \underline{h}^{00} \partial^I L^{J_2'} \partial_t\partial_t \phi, \quad \underline{h}^{00} \partial_\alpha\partial_\beta \partial^I L^{J'} \phi,
\end{aligned} \tag{4.16}$$

where $I_1 + I_2 = I, J_1 + J_2 = J$ with $|I_1| \geq 1$, $J_1' + J_2' = J$ with $|J_1'| \geq 1$ and $|J'| < |J|$ and $|I_3| + |I_4| \leq |I|, |J_3| + |J_4| \leq |J|$. In our analysis of the commutator estimates, we will make use of the decompositions (4.14) and (4.16).

4.4 Estimates based on commutators and homogeneity

Let u be a smooth function defined in $\mathcal{K}$ and vanishing near the boundary $\{r = t-1\}$. In view of $\underline{\partial}_a = t^{-1}L_a$, we have

$$\partial^I L^J \underline{\partial}_a u = \partial^I L^J (t^{-1} L_a u) = \sum_{\substack{I_1+I_2=I \\ J_1+J_2=J}} \partial^{I_1} L^{J_1}(t^{-1}) \partial^{I_2} L^{J_2} L_a u.$$

Since t^{-1} is a smooth and homogeneous coefficient of degree -1, we have

$$\big|\partial^I L^J \underline{\partial}_a u\big| \leq C t^{-1} \sum_{\substack{|I'|\leq|I| \\ |J'|\leq|J|}} \big|\partial^{I'} L^{J'} L_a u\big|. \tag{4.17}$$

As a direct application, for instance we have

$$\begin{aligned}
&\big|\partial^I L^J \underline{\partial}_a \underline{\partial}_\nu u\big| \\
&\leq C t^{-1} \sum_{\substack{|I'|\leq|I| \\ |J'|\leq|J|}} \big|\partial^{I'} L^{J'} L_a \underline{\partial}_\nu u\big| = C t^{-1} \sum_{\substack{|I'|\leq|I| \\ |J'|\leq|J|}} \big|\partial^{I'} L^{J'} L_a \big(\Phi_\nu^{\nu'} \partial_{\nu'} u\big)\big|.
\end{aligned}$$

The function $\Phi_\nu^{\nu'}$ is smooth and homogeneous of degree 0, so that

$$\big|\partial^I L^J \underline{\partial}_a \underline{\partial}_\nu u\big| \leq C(I,J) t^{-1} \sum_{\substack{\gamma, |I'|\leq|I| \\ |J'|\leq|J|}} \big|\partial^{I'} L^{J'} L_a \partial_\gamma u\big|. \tag{4.18}$$

A similar argument holds for

$$\big|\partial^I L^J \underline{\partial}_\nu \underline{\partial}_a u\big| \leq C(I,J) t^{-1} \sum_{\substack{\gamma, a, |I'|\leq|I| \\ |J'|\leq|J|}} \big|\partial^{I'} L^{J'} L_a \partial_\gamma u\big|. \tag{4.19}$$

Furthermore, when there are two "good" derivatives, we consider

$$\begin{aligned}
&\partial^I L^J (\underline{\partial}_a \underline{\partial}_b u) = \partial^I L^J \big(t^{-1} L_a (t^{-1} L_b) u\big) \\
&= \partial^I L^J (t^{-2} L_a L_b u) + \partial^I L^J \big(t^{-1} L_a (t^{-1}) u\big) \\
&= \sum_{\substack{I_1+I_2=I \\ J_1+J_2=J}} \partial^{I_1} L^{J_1}(t^{-2}) \partial^{I_2} L^{J_2} L_a L_b u + \sum_{\substack{I_1+I_2=I \\ J_1+J_2=J}} \partial^{I_1} L^{J_1}\big(t^{-1} L_a (t^{-1})\big) \partial^{I_2} L^{J_2} L_a u,
\end{aligned}$$

and we find

$$\begin{aligned}
\big|\partial^I L^J (\underline{\partial}_a \underline{\partial}_b u)\big| &= \big|\partial^I L^J \big(t^{-1} L_a (t^{-1} L_b) u\big)\big| \\
&\leq C t^{-2} \sum_{\substack{|I'|\leq|I| \\ |J'|\leq|J|}} \big|\partial^{I'} L^{J'} L_a L_b u\big| + C t^{-2} \sum_{\substack{|I'|\leq|I| \\ |J'|\leq|J|}} \big|\partial^{I'} L^{J'} L_b u\big|.
\end{aligned} \tag{4.20}$$

4.5 Basic structure of the quasi-null terms

In this section we consider the quasi-null terms $P_{\alpha\beta}$ and emphasize some important properties:
1. The expression $P_{\alpha\beta}$ is a 2-tensor and this tensorial structure plays a role in our analysis.
2. In explicit form, it reads

$$P_{\alpha\beta} = \frac{1}{4} g^{\gamma\gamma'} g^{\delta\delta'} \partial_\alpha h_{\gamma\delta} \partial_\beta h_{\gamma'\delta'} - \frac{1}{2} g^{\gamma\gamma'} g^{\delta\delta'} \partial_\alpha h_{\gamma\gamma'} \partial_\beta h_{\delta\delta'},$$

and, in the semi-hyperboloidal frame,

$$\underline{P}_{\alpha\beta} = \frac{1}{4} g^{\gamma\gamma'} g^{\delta\delta'} \underline{\partial}_\alpha h_{\gamma\delta} \underline{\partial}_\beta h_{\gamma'\delta'} - \frac{1}{2} g^{\gamma\gamma'} g^{\delta\delta'} \underline{\partial}_\alpha h_{\gamma\gamma'} \underline{\partial}_\beta h_{\delta\delta'},$$

so the only term to be concerned about is the 00-component.

$$\begin{aligned}\underline{P}_{00} &= \frac{1}{4} g^{\gamma\gamma'} g^{\delta\delta'} \partial_t h_{\gamma\delta} \partial_t h_{\gamma'\delta'} - \frac{1}{2} g^{\gamma\gamma'} g^{\delta\delta'} \partial_t h_{\gamma\gamma'} \partial_t h_{\delta\delta'} \\ &= \frac{1}{4} \underline{g}^{\gamma\gamma'} \underline{g}^{\delta\delta'} \partial_t \underline{h}_{\gamma\delta} \partial_t \underline{h}_{\gamma'\delta'} - \frac{1}{2} \underline{g}^{\gamma\gamma'} \underline{g}^{\delta\delta'} \partial_t \underline{h}_{\gamma\gamma'} \partial_t \underline{h}_{\delta\delta'} + Com(0,0).\end{aligned}$$

Here $Com(0,0)$ represents the commutator terms:

$$\begin{aligned}Com(0,0) = {} & \frac{1}{4} g^{\gamma\gamma'} g^{\delta\delta'} \underline{h}_{\gamma''\delta''} \partial_t \big(\Psi_\gamma^{\gamma''} \Psi_\delta^{\delta''}\big) \partial_t \big(\Psi_{\gamma'}^{\gamma'''} \Psi_{\delta'}^{\delta'''}\big) \underline{h}_{\gamma'''\delta'''} \\ & + \frac{1}{4} g^{\gamma\gamma'} g^{\delta\delta'} \Psi_\gamma^{\gamma''} \Psi_\delta^{\delta''} \partial_t \underline{h}_{\gamma''\delta''} \partial_t \big(\Psi_{\gamma'}^{\gamma'''} \Psi_{\delta'}^{\delta'''}\big) \underline{h}_{\gamma'''\delta'''} \\ & + \frac{1}{4} g^{\gamma\gamma'} g^{\delta\delta'} \partial_t \big(\Psi_\gamma^{\gamma''} \Psi_\delta^{\delta''}\big) \underline{h}_{\gamma''\delta''} \Psi_{\gamma'}^{\gamma''} \Psi_{\delta'}^{\delta''} \partial_t \underline{h}_{\gamma'''\delta'''} \\ & - \frac{1}{2} g^{\gamma\gamma'} g^{\delta\delta'} \partial_t \big(\Psi_\gamma^{\gamma''} \Psi_\delta^{\delta''}\big) \underline{h}_{\gamma''\gamma'''} \partial_t \big(\Psi_{\gamma'}^{\gamma'''} \Psi_{\delta'}^{\delta'''}\big) \underline{h}_{\delta''\delta'''} \\ & - \frac{1}{2} g^{\gamma\gamma'} g^{\delta\delta'} \Psi_\gamma^{\gamma''} \Psi_\delta^{\delta''} \partial_t \underline{h}_{\gamma''\gamma'''} \partial_t \big(\Psi_{\gamma'}^{\gamma'''} \Psi_{\delta'}^{\delta'''}\big) \underline{h}_{\delta''\delta'''} \\ & - \frac{1}{2} g^{\gamma\gamma'} g^{\delta\delta'} \partial_t \big(\Psi_\gamma^{\gamma''} \Psi_\delta^{\delta''}\big) \underline{h}_{\gamma''\gamma'''} \Psi_{\gamma'}^{\gamma''} \Psi_{\delta'}^{\delta''} \partial_t \underline{h}_{\delta''\delta'''}.\end{aligned}$$

We see that

$$\begin{aligned}\underline{P}_{00} &= \frac{1}{4} \underline{g}^{\gamma\gamma'} \underline{g}^{\delta\delta'} \partial_t \underline{h}_{\gamma\delta} \partial_t \underline{h}_{\gamma'\delta'} - \frac{1}{2} \underline{g}^{\gamma\gamma'} \underline{g}^{\delta\delta'} \partial_t \underline{h}_{\gamma\gamma'} \partial_t \underline{h}_{\delta\delta'} + Com(0,0) \\ &= \frac{1}{4} \underline{m}^{\gamma\gamma'} \underline{m}^{\delta\delta'} \partial_t \underline{h}_{\gamma\delta} \partial_t \underline{h}_{\gamma'\delta'} - \frac{1}{2} \underline{g}^{\gamma\gamma'} \underline{g}^{\delta\delta'} \partial_t \underline{h}_{\gamma\gamma'} \partial_t \underline{h}_{\delta\delta'} \\ &\quad + Com(0,0) + Cub(0,0).\end{aligned}$$

Here the terms $Cub(0,0)$ stands for the high-order terms:

$$\begin{aligned}Cub(0,0) = {} & \frac{1}{4} \underline{h}^{\gamma\gamma'} \underline{m}^{\delta\delta'} \partial_t \underline{h}_{\gamma\delta} \partial_t \underline{h}_{\gamma'\delta'} + \frac{1}{4} \underline{m}^{\gamma\gamma'} \underline{h}_{\gamma\delta'} \partial_t \underline{h}_{\gamma\delta} \partial_t \underline{h}_{\gamma'\delta'} \\ & + \frac{1}{4} \underline{h}^{\gamma\gamma'} \underline{h}^{\delta\delta'} \partial_t \underline{h}_{\gamma\delta'} \partial_t \underline{h}_{\gamma\delta'}.\end{aligned}$$

We summarize our conclusion.

Lemma 4.4 (Structure of the quasi-null terms). *The quasi-null term* $\underline{P}_{00}$ *are linear combinations of the following terms with smooth and homogeneous coefficients of degree* ≤ 0*:*

$$\begin{aligned}&GQS_h(0,0), \quad Cub(0,0), \quad Com(0,0), \quad \underline{g}^{\gamma\gamma'}\underline{g}^{\delta\delta'}\partial_t\underline{h}_{\gamma\gamma'}\partial_t\underline{h}_{\delta\delta'},\\ &\underline{m}^{\gamma\gamma'}\underline{m}^{\delta\delta'}\partial_t\underline{h}_{\gamma\delta}\partial_t\underline{h}_{\gamma'\delta'}.\end{aligned} \tag{4.21}$$

The quasi-null term $\underline{P}_{a\beta}$ *are linear combinations of* $GQS_h(0,0)$ *and* $Cub(0,0)$ *terms.*

So, the only problematic terms in $P_{\alpha\beta}$ are $\underline{g}^{\gamma\gamma'}\underline{g}^{\delta\delta'}\partial_t\underline{h}_{\gamma\gamma'}\partial_t\underline{h}_{\delta\delta'}$ and $\underline{m}^{\gamma\gamma'}\underline{m}^{\delta\delta'}\partial_t\underline{h}_{\gamma\delta}\partial_t\underline{h}_{\gamma'\delta'}$. They will be controlled by using the wave gauge condition.

4.6 Metric components in the semi-hyperboloidal frame

In this subsection, we derive the equation satisfied by the metric components within the semi-hyperboloidal frame. To do so, we need the identity

$$\widetilde{\Box}_g(uv) = u\widetilde{\Box}_g v + v\widetilde{\Box}_g u + 2g^{\alpha\beta}\partial_\alpha u\partial_\beta v.$$

Then, we have

$$\begin{aligned}\widetilde{\Box}_g\underline{h}_{\alpha\beta} &= \widetilde{\Box}_g\left(\Phi_\alpha^{\alpha'}\Phi_\beta^{\beta'}h_{\alpha'\beta'}\right)\\ &= \Phi_\alpha^{\alpha'}\Phi_\beta^{\beta'}\widetilde{\Box}_g h_{\alpha'\beta'} + 2g^{\mu\nu}\partial_\mu\left(\Phi_\alpha^{\alpha'}\Phi_\beta^{\beta'}\right)\partial_\nu h_{\alpha'\beta'} + h_{\alpha'\beta'}\widetilde{\Box}_g\left(\Phi_\alpha^{\alpha'}\Phi_\beta^{\beta'}\right).\end{aligned}$$

Then we calculate explicitly the correction terms concerning the derivatives of $\Phi_\alpha^{\alpha'}\Phi_\beta^{\beta'}$:

- Case $\alpha = \beta = 0$:

$$\Phi_0^0\Phi_0^0 = 1, \text{ the other ones vanish,}$$

$$\Box\left(\Phi_0^{\alpha'}\Phi_0^{\beta'}\right) = 0, \quad \partial\left(\Phi_0^0\Phi_0^0\right) = 0.$$

- Case $\alpha = a > 0, \beta = 0$:

$$\Phi_a^0\Phi_0^0 = x^a/t, \quad \Phi_a^a\Phi_0^0 = 1,$$

and

$$\Box\left(\Phi_a^0\Phi_0^0\right) = -\frac{2x^a}{t^3}, \quad \partial_t\left(\Phi_a^0\Phi_0^0\right) = -\frac{x^a}{t^2}, \quad \partial_a\left(\Phi_a^0\Phi_0^0\right) = \frac{1}{t}.$$

- Case $\alpha = a > 0, \beta = b > 0$:

$$\Phi_a^0 \Phi_b^0 = x^a x^b / t^2, \quad \Phi_a^0 \Phi_b^b = x^a/t, \quad \Phi_a^a \Phi_b^b = 1,$$

and

$$\begin{aligned}
\Box(\Phi_a^0 \Phi_b^0) &= -\frac{6x^a x^b}{t^4} + \frac{2\delta_{ab}}{t^2}, \quad \partial_t(\Phi_a^0 \Phi_b^0) = -\frac{2x^a x^b}{t^3},\\
\partial_c(\Phi_a^0 \Phi_b^0) &= \frac{\delta_{ca} x^b + \delta_{cb} x^a}{t^2},\\
\Box(\Phi_a^0 \Phi_b^b) &= -\frac{2x^a}{t^3}, \quad \partial_t(\Phi_a^0 \Phi_b^b) = -\frac{x^a}{t^2}, \quad \partial_a(\Phi_a^0 \Phi_b^b) = \frac{1}{t},
\end{aligned}$$

while the other ones vanish.

Then we calculate the remaining terms (up to second-order):

$$\begin{aligned}
\widetilde{\Box}_g \underline{h}_{00} &= \Phi_0^{\alpha'} \Phi_0^{\beta'} Q_{\alpha'\beta'} + \underline{P}_{00} - 16\pi \partial_t \phi \partial_t \phi - 8\pi c^2 \underline{m}_{00} \phi^2 + Cub(0,0),\\
\widetilde{\Box}_g \underline{h}_{0a} &= \Phi_0^{\alpha'} \Phi_a^{\beta'} Q_{\alpha'\beta'} + \underline{P}_{0a} - 16\pi \underline{\partial}_a \phi \partial_t \phi - 8\pi c^2 \underline{m}_{a0} \phi^2 + \frac{2}{t} \underline{\partial}_a h_{00}\\
&\quad - \frac{2x^a}{t^3} h_{00} + Cub(0,0),\\
\widetilde{\Box}_g \underline{h}_{aa} &= \Phi_a^{\alpha'} \Phi_a^{\beta'} Q_{\alpha'\beta'} + \underline{P}_{aa} - 16\pi \underline{\partial}_a \phi \underline{\partial}_a \phi - 8\pi c^2 \underline{m}_{aa} \phi^2,\\
&\quad + \frac{4x^a}{t^2} \underline{\partial}_a h_{00} + \frac{4}{t} \underline{\partial}_a h_{0a} - \frac{4x^a}{t^3} h_{0a} + \left(\frac{2}{t^2} - \frac{6|x^a|^2}{t^4}\right) h_{00} + Cub(0,0),\\
\widetilde{\Box}_g \underline{h}_{ab} &= \Phi_a^{\alpha'} \Phi_b^{\beta'} Q_{\alpha'\beta'} + \underline{P}_{ab} - 16\pi \underline{\partial}_a \phi \underline{\partial}_b \phi - 8\pi c^2 \underline{m}_{ab} \phi^2,\\
&\quad + \frac{2x^b}{t^2} \underline{\partial}_a h_{00} + \frac{2x^a}{t^2} \underline{\partial}_b h_{00} + \frac{2}{t} \underline{\partial}_a h_{0b} + \frac{2}{t} \underline{\partial}_b h_{0a} - \frac{6x^a x^b}{t^4} h_{00}\\
&\quad - \frac{2x^a}{t^3} h_{0b} - \frac{2x^b}{t^3} h_{0a} + Cub(0,0)\\
&\quad (a \neq b).
\end{aligned}$$

The most important point is that for the components $\underline{h}_{\alpha\beta}$, the quasi-null terms $P_{\alpha\beta}$ become *null terms*. This tensorial structure will lead us to the fact that these metric components do have better decay rate compared to $\underline{h}_{00}$. In Section 9, these equations will be used to derive sharp decay estimates for these components. For clarity, we state the following conclusion:

$$
\begin{aligned}
\widetilde{\Box}_g \underline{h}_{0a} &= \frac{2}{t}\underline{\partial}_a h_{00} - \frac{2x^a}{t^3} h_{00} + GQS_h(0,0) + GQS_\phi(0,0) + Cub(0,0),\\
\widetilde{\Box}_g \underline{h}_{aa} &= \frac{4x^a}{t^2}\underline{\partial}_a h_{00} + \left(\frac{2}{t^2} - \frac{6|x^a|^2}{t^4}\right) h_{00} + \frac{4}{t}\underline{\partial}_a h_{0a} - \frac{4x^a}{t^3} h_{0a}\\
&\quad + GQS_h(0,0) + GQS_\phi(0,0) + Cub(0,0),\\
\widetilde{\Box}_g \underline{h}_{ab} &= \frac{2x^b}{t^2}\underline{\partial}_a h_{00} + \frac{2x^a}{t^2}\underline{\partial}_b h_{00} - \frac{6x^a x^b}{t^4} h_{00} + \frac{2}{t}\underline{\partial}_a h_{0b}\\
&\quad - \frac{2x^a}{t^3} h_{0b} + \frac{2}{t}\underline{\partial}_a h_{0a} - \frac{2x^b}{t^3} h_{0a}\\
&\quad + GQS_h(0,0) + GQS_\phi(0,0) + Cub(0,0).
\end{aligned}
\tag{4.22}
$$

4.7 Wave gauge condition in the semi-hyperboloidal frame

Our objective in the rest of this section is to establish some estimates based on the wave condition $g^{\alpha\beta}\Gamma^\gamma_{\alpha\beta} = 0$, which is equivalent to saying

$$
g_{\beta\gamma}\partial_\alpha g^{\alpha\beta} = \frac{1}{2} g_{\alpha\beta}\partial_\gamma g^{\alpha\beta}. \tag{4.23}
$$

We have introduced

$$
\begin{aligned}
h^{\alpha\beta} &= g^{\alpha\beta} - m^{\alpha\beta}, \quad h_{\alpha\beta} = g_{\alpha\beta} - m_{\alpha\beta},\\
\underline{h}^{\alpha\beta} &= \underline{g}^{\alpha\beta} - \underline{m}^{\alpha\beta}, \quad \underline{h}_{\alpha\beta} = \underline{g}_{\alpha\beta} - \underline{m}_{\alpha\beta},
\end{aligned}
\tag{4.24}
$$

in which $\underline{h}^{\alpha\beta} = h^{\alpha'\beta'}\Psi^\alpha_{\alpha'}\Psi^\beta_{\beta'}$ and $\underline{h}_{\alpha\beta} = h_{\alpha'\beta'}\Phi^{\alpha'}_\alpha\Phi^{\beta'}_\beta$.

Lemma 4.5. *Let $(g_{\alpha\beta})$ be a metric satisfying the wave gauge condition (4.23). Then $\partial_t \underline{h}^{00}$ is a linear combination of the following terms with smooth and homogeneous coefficients of degree ≤ 0:*

$$
(s/t)^2\partial_\alpha \underline{h}^{\beta\gamma}, \quad \underline{\partial}_a \underline{h}^{\beta\gamma}, \quad t^{-1}\underline{h}^{\alpha\beta}, \quad \underline{h}^{\alpha\beta}\partial_\gamma \underline{h}^{\alpha'\beta'}, \quad t^{-1} h_{\alpha\beta}\underline{h}^{\alpha'\beta'}. \tag{4.25}
$$

Proof. The wave gauge condition (4.23) can be written in the semi-hyperboloidal frame as

$$
\underline{g}_{\beta\gamma}\underline{\partial}_\alpha \underline{h}^{\alpha\beta} + g_{\beta'\gamma'}\Phi^{\gamma'}_\gamma \underline{h}^{\alpha\beta}\partial_{\alpha'}(\Phi^{\alpha'}_\alpha \Phi^{\beta'}_\beta) = \frac{1}{2}\underline{g}_{\alpha\beta}\underline{\partial}_\gamma \underline{h}^{\alpha\beta} + \frac{1}{2} g_{\alpha\beta}\underline{h}^{\alpha'\beta'}\underline{\partial}_\gamma(\Phi^\alpha_{\alpha'}\Phi^\beta_{\beta'}). \tag{4.26}
$$

This leads us to

$$
\begin{aligned}
\underline{m}_{\beta\gamma}\underline{\partial}_\alpha \underline{h}^{\alpha\beta} = {} & \frac{1}{2}\underline{g}_{\alpha\beta}\underline{\partial}_\gamma \underline{h}^{\alpha\beta} + \frac{1}{2} g_{\alpha\beta}\underline{h}^{\alpha'\beta'}\underline{\partial}_\gamma(\Phi^\alpha_{\alpha'}\Phi^\beta_{\beta'})\\
& - g_{\beta'\gamma'}\Phi^{\gamma'}_\gamma \underline{h}^{\alpha\beta}\partial_{\alpha'}(\Phi^{\alpha'}_\alpha \Phi^{\beta'}_\beta) - \underline{h}_{\beta\gamma}\underline{\partial}_\alpha \underline{h}^{\alpha\beta}.
\end{aligned}
\tag{4.27}
$$

Taking $\gamma = c = 1,2,3$, we analyze the left-hand side and observe that

$$\underline{m}_{\beta c}\underline{\partial}_{\alpha}\underline{h}^{\alpha\beta} = \underline{m}_{0c}\underline{\partial}_0\underline{h}^{00} + \underline{m}_{bc}\underline{\partial}_0\underline{h}^{0b} + \underline{m}_{\beta c}\underline{\partial}_a\underline{h}^{a\beta},$$

which leads us to $\underline{m}_{0c}\underline{\partial}_0\underline{h}^{00} = \underline{m}_{\beta c}\underline{\partial}_{\alpha}\underline{h}^{\alpha\beta} - \underline{m}_{bc}\underline{\partial}_0\underline{h}^{0b} - \underline{m}_{\beta c}\underline{\partial}_a\underline{h}^{a\beta}$, so that

$$\underline{m}^{0c}\underline{m}_{0c}\underline{\partial}_0\underline{h}^{00} = \underline{m}^{0c}\underline{m}_{\beta c}\underline{\partial}_{\alpha}\underline{h}^{\alpha\beta} - \underline{m}^{0c}\underline{m}_{bc}\underline{\partial}_0\underline{h}^{0b} - \underline{m}^{0c}\underline{m}_{\beta c}\underline{\partial}_a\underline{h}^{a\beta'}.$$

An explicit calculation shows that $\underline{m}^{0c}\underline{m}_{0c} = \frac{r^2}{t^2}$, $\underline{m}^{0c}\underline{m}_{bc} = -(s/t)^2(x^b/t)$ and thus

$$(r/t)^2\underline{\partial}_0\underline{h}^{00} = \underline{m}^{0c}\underline{m}_{\beta c}\underline{\partial}_{\alpha}\underline{h}^{\alpha\beta} + (s/t)^2\sum_b(x^b/t)\underline{\partial}_0\underline{h}^{0b} - \underline{m}^{0c}\underline{m}_{\beta c}\underline{\partial}_a\underline{h}^{a\beta'}. \tag{4.28}$$

Combining (4.27) and (4.28), we find

$$\begin{aligned}(r/t)^2\underline{\partial}_0\underline{h}^{00} =&(s/t)^2\sum_b(x^b/t)\underline{\partial}_0\underline{h}^{0b} - \underline{m}^{0c}\underline{m}_{\beta c}\underline{\partial}_a\underline{h}^{a\beta'}\\ &+\underline{m}^{0c}\Big(\frac{1}{2}\underline{g}_{\alpha\beta}\underline{\partial}_c\underline{h}^{\alpha\beta} + \frac{1}{2}g_{\alpha\beta}\underline{h}^{\alpha'\beta'}\underline{\partial}_c(\Phi^{\alpha}_{\alpha'}\Phi^{\beta}_{\beta'})\\ &- g_{\beta'\gamma'}\Phi^{\gamma'}_c\underline{h}^{\alpha\beta}\partial_{\alpha'}(\Phi^{\alpha'}_{\alpha}\Phi^{\beta'}_{\beta}) - \underline{h}_{\beta c}\underline{\partial}_{\alpha}\underline{h}^{\alpha\beta}\Big),\end{aligned} \tag{4.29}$$

which gives us the terms in (4.25). □

We now proceed by deriving some estimates based on the wave gauge condition. For convenience, we introduce the notation (with $c = 1,2,3$)

$$|\underline{h}| := \max_{\alpha,\beta}|\underline{h}_{\alpha\beta}|, \qquad |\partial\underline{h}| := \max_{\alpha,\beta,\gamma}|\partial_{\gamma}\underline{h}_{\alpha\beta}|, \qquad |\underline{\partial h}| := \max_{c,\alpha,\beta}|\underline{\partial}_c\underline{h}_{\alpha\beta}|.$$

Observe that $|\underline{\partial h}|$ contains only the "good" derivatives of $\underline{h}_{\alpha\beta}$. When $|\partial\underline{h}|$ and $|\underline{h}|$ are supposed to be small enough, and, the rest of this section, we express the corresponding bound in the form $\varepsilon_w \leq 1$, the algebraic relation between $\underline{h}^{\alpha\beta}$ and $\underline{h}_{\alpha\beta}$ leads us to the following basic estimates:

$$\max_{\alpha,\beta}|\underline{h}^{\alpha\beta}| \leq C|\underline{h}|, \qquad \max_{\alpha,\beta,\gamma}|\partial_{\gamma}\underline{h}^{\alpha\beta}| \leq C|\partial\underline{h}|, \qquad \max_{c,\alpha,\beta}|\underline{\partial}_c\underline{h}^{\alpha\beta}| \leq C|\underline{\partial h}|. \tag{4.30}$$

With the above preparation, the following estimate is immediate from Lemma 4.5.

Lemma 4.6 (Zero-order wave coordinate estimate). *Let $g^{\alpha\beta} = m^{\alpha\beta} + h^{\alpha\beta}$ be a metric satisfying the wave gauge condition (4.23). We suppose furthermore that $|\partial\underline{h}|$ and $|\underline{h}|$ are small enough so (4.30) hold. Then the following estimate holds:*

$$|\underline{\partial}_t\underline{h}^{00}| \leq C(s/t)^2|\partial\underline{h}| + C|\underline{\partial h}| + Ct^{-1}|\underline{h}| + C|\partial\underline{h}|\,|\underline{h}|. \tag{4.31}$$

The interest of this estimate is as follows: the "bad" derivative of $\underline{h}^{00}$ is bounded by the "good" derivatives arising in the right-hand side of (4.31). Of course, the "bad" term $|\partial \underline{h}|$ still arise, but it is multiplied by the factor $(s/t)^2$ which provides us with extra decay and turns this term into a "good" term.

Lemma 4.7 (k-order wave coordinate estimates). *Let $g^{\alpha\beta} = m^{\alpha\beta} + h^{\alpha\beta}$ be a smooth metric satisfying the wave gauge condition (4.23). We suppose furthermore that for a product $\partial^I L^J$ with $|I| + |J| \leq N$, $|\partial\partial^I L^J \underline{h}|$ and $|\partial^I L^J \underline{h}|$ are small enough so that the following bounds hold:* $\max_{\alpha,\beta} |\partial^I L^J \underline{h}^{\alpha\beta}| \leq C|\partial^I L^J \underline{h}|$, $\max_{\alpha,\beta,\gamma} |\partial_\gamma \partial^I L^J \underline{h}^{\alpha\beta}| \leq C|\partial\partial^I L^J \underline{h}|$, *and* $\max_{c,\alpha,\beta} |\underline{\partial}_c \partial^I L^J \underline{h}^{\alpha\beta}| \leq C|\underline{\partial}\partial^I L^J \underline{h}|$. *Then the following estimate holds:*

$$
\begin{aligned}
&|\partial^I L^J \partial_t \underline{h}^{00}| + |\partial_t \partial^I L^J \underline{h}^{00}| \\
&\leq C \sum_{\substack{|I'|+|J'|\leq|I|+|J| \\ |J'|\leq|J|}} \left((s/t)^2 |\partial\partial^{I'} L^{J'} \underline{h}| + |\partial^{I'} L^{J'} \underline{\partial h}| + t^{-1}|\partial^{I'} L^{J'} \underline{h}|\right) \\
&\quad + C \sum_{\substack{|I_1|+|I_2|\leq|I| \\ |J_1|+|J_2|\leq|J|}} |\partial^{I_1} L^{J_1} \underline{h}|\, |\partial\partial^{I_2} L^{J_2} \underline{h}|.
\end{aligned}
\tag{4.32}
$$

Proof. This result is also a direct consequence of Lemma 4.5. We derive the expression of $\partial_t \underline{h}^{00}$ which is a linear combination of the terms in (4.25) with smooth and homogeneous coefficients of degree ≤ 0. So, $\partial^I L^J \partial_t \underline{h}^{00}$ is again a linear combination of the following terms with smooth and homogeneous coefficients of degree $\leq |I|$ (since $\partial^I L^J$ acts on a 0-homogeneous function gives a $|I|$-homogeneous function):
$\partial^{I'} L^{J'}\left((s/t)^2 \partial_\alpha \underline{h}^{\beta\gamma}\right)$, $\partial^{I'} L^{J'}\left(\underline{\partial}_a \underline{h}^{\beta\gamma}\right)$, $t^{-1}\partial^{I'} L^{J'}\left(\underline{h}^{\alpha\beta}\right)$, $\partial^{I'} L^{J'}\left(\underline{h}^{\alpha\beta} \partial_\gamma \underline{h}^{\alpha'\beta'}\right)$, $t^{-1}\partial^{I'} L^{J'}\left(h_{\alpha\beta} \underline{h}^{\alpha'\beta'}\right)$
with $|I'| \leq |I|$ and $|J'| \leq |J|$. We observe that

$$
|\partial^{I'} L^{J'}\left((s/t)^2 \partial_\alpha \underline{h}^{\beta\gamma}\right)| \leq C(s/t)^2 \sum_{\substack{|I''|\leq|I'| \\ |J''|\leq|J'|}} |\partial^{I''} L^{J''}\left(\partial_\alpha \underline{h}^{\beta\gamma}\right)|.
$$

The second, fourth, and last terms are to be bounded by the commutator estimates in Lemma 3.15. The estimate for $\partial_t \partial^I L^J \underline{h}^{00}$ is deduced from (4.32) and the commutator estimates. □

4.8 Revisiting the structure of the quasi-null terms

In this section, we consider the estimates on quasi-null terms $P_{\alpha\beta}$ together with the wave gauge condition and we use wave coordinate estimates. We treat first the term $\underline{g}^{\alpha\alpha'}\partial_t\underline{g}_{\beta\beta'}$ and formulate the wave gauge condition in the form:

$$g^{\alpha\beta}\partial_\alpha h_{\beta\gamma} = \frac{1}{2}g^{\alpha\beta}\partial_\gamma h_{\alpha\beta}. \tag{4.33}$$

Lemma 4.8. *There exists a positive constant $\varepsilon_w \geq 0$ such that if $|h|+|\partial h| \leq \varepsilon_w$, and the wave gauge condition (4.33) holds, then the quasi-null term $\underline{g}^{\alpha\alpha'}\underline{g}^{\beta\beta'}\partial_t\underline{g}_{\alpha\alpha'}\partial_t\underline{g}_{\beta\beta'}$ is a linear combination of terms*

$$GQS_h(0,0), \quad Com(0,0), \quad Cub(0,0), \quad \underline{g}^{0a}\underline{\partial}_0\underline{g}_{0a}\underline{g}^{0b}\underline{\partial}_0\underline{g}_{0b} \tag{4.34}$$

with smooth and homogeneous coefficients of degree ≤ 0.

Proof. The relation (4.33) can be written in the semi-hyperboloidal frame in the form:

$$\begin{aligned}&\underline{g}^{\alpha\beta}\underline{\partial}_\alpha\underline{h}_{\beta\gamma} + \Phi_\gamma^{\gamma'}\underline{g}^{\alpha\beta}\partial_\alpha\left(\Psi_\beta^{\beta'}\Psi_{\gamma'}^{\gamma''}\right)\underline{h}_{\beta'\gamma''}\\ &= \frac{1}{2}\underline{g}^{\alpha\beta}\underline{\partial}_\gamma\underline{h}_{\alpha\beta} + \frac{1}{2}\underline{g}^{\alpha\beta}\underline{\partial}_\gamma\left(\Psi_\alpha^{\alpha'}\Psi_\beta^{\beta'}\right)\underline{h}_{\alpha'\beta'}.\end{aligned} \tag{4.35}$$

We fix $\gamma = 0$ and see that

$$\underline{g}^{\alpha\beta}\partial_t\underline{h}_{\alpha\beta} = 2\underline{g}^{\alpha\beta}\underline{\partial}_\alpha\underline{h}_{0\beta} + 2\Phi_0^{\gamma'}\underline{g}^{\alpha\beta}\partial_\alpha\left(\Psi_\beta^{\beta'}\Psi_{\gamma'}^{\gamma''}\right)\underline{h}_{\beta'\gamma''} - \underline{g}^{\alpha\beta}\partial_t\left(\Psi_\alpha^{\alpha'}\Psi_\beta^{\beta'}\right)\underline{h}_{\alpha'\beta'}.$$

This identity can be written as

$$\begin{aligned}\underline{g}^{\alpha\beta}\partial_t\underline{h}_{\alpha\beta} =& 2\underline{m}^{\alpha\beta}\underline{\partial}_\alpha\underline{h}_{\beta 0} + 2\underline{h}^{\alpha\beta}\underline{\partial}_\alpha\underline{h}_{\beta 0}\\ &+ 2\Phi_0^{\gamma'}\underline{m}^{\alpha\beta}\partial_\alpha\left(\Psi_\beta^{\beta'}\Psi_{\gamma'}^{\gamma''}\right)\underline{h}_{\beta'\gamma''} - \underline{m}^{\alpha\beta}\partial_t\left(\Psi_\alpha^{\alpha'}\Psi_\beta^{\beta'}\right)\underline{h}_{\alpha'\beta'}\\ &+ 2\Phi_0^{\gamma'}\underline{h}^{\alpha\beta}\partial_\alpha\left(\Psi_\beta^{\beta'}\Psi_{\gamma'}^{\gamma''}\right)\underline{h}_{\beta'\gamma''} - \underline{h}^{\alpha\beta}\partial_t\left(\Psi_\alpha^{\alpha'}\Psi_\beta^{\beta'}\right)\underline{h}_{\alpha'\beta'}.\end{aligned} \tag{4.36}$$

In the right-hand side, except for the first term, we have at least quadratic terms or terms containing an extra decay factor such as $\partial_\alpha\left(\Psi_\beta^{\beta'}\Psi_\gamma^{\gamma''}\right)$. So, we see that in $\underline{g}^{\alpha\alpha'}\underline{g}^{\beta\beta'}\partial_t\underline{g}_{\alpha\alpha'}\partial_t\underline{g}_{\beta\beta'}$ the only term to be concerned about is

$$4\underline{m}^{\alpha\alpha'}\underline{m}^{\beta\beta'}\underline{\partial}_\alpha\underline{h}_{\alpha'0}\underline{\partial}_\beta\underline{h}_{\beta'0}.$$

The remaining terms are quadratic in $\underline{h}^{\alpha\beta}$, $\underline{h}_{\alpha\beta}$ or linear terms on $\underline{h}_{\alpha\beta}$ with decreasing coefficients such as $\partial_\alpha\left(\Psi_\beta^{\beta'}\Psi_\gamma^{\gamma''}\right)$. Then we also see that when $|\underline{h}|$ sufficiently small, $\underline{h}^{\alpha\beta}$ can be expressed as a power series of $\underline{h}_{\alpha\beta}$

(without zero order), which is itself a linear combination of $h_{\alpha\beta}$ with smooth and homogeneous coefficients of degree ≤ 0. So, when $|h|$ sufficiently small, $\underline{h}^{\alpha\beta}$ can be expressed as a power series of $h_{\alpha\beta}$ (without 0 order) with smooth and homogeneous coefficients of degree ≤ 0. We conclude that in the product $\underline{g}^{\alpha\alpha'}\underline{g}^{\beta\beta'}\partial_t\underline{g}_{\alpha\alpha'}\partial_t\underline{g}_{\beta\beta'}$, the remaining terms apart from $4\underline{m}^{\alpha\alpha'}\underline{m}^{\beta\beta'}\underline{\partial}_\alpha\underline{h}_{\alpha'0}\underline{\partial}_\beta\underline{h}_{\beta'0}$ are contained in $Cub(0,0)$ or $Com(0,0)$.

We focus on the term $4\underline{m}^{\alpha\alpha'}\underline{m}^{\beta\beta'}\underline{\partial}_\alpha\underline{h}_{\alpha'0}\underline{\partial}_\beta\underline{h}_{\beta'0}$. We have

$$
\begin{aligned}
&4\big(\underline{m}^{\alpha\alpha'}\underline{\partial}_\alpha\underline{h}_{\alpha'0}\big)\big(\underline{m}^{\beta\beta'}\underline{\partial}_\beta\underline{h}_{\beta'0}\big)\\
&= 4\big(\underline{m}^{a\alpha'}\underline{\partial}_a\underline{h}_{\alpha'0} + \underline{m}^{00}\underline{\partial}_0\underline{h}_{00} + \underline{m}^{0a'}\underline{\partial}_0\underline{h}_{0a'}\big)\\
&\quad \times \big(\underline{m}^{b\beta'}\underline{\partial}_b\underline{h}_{\beta'0} + \underline{m}^{00}\underline{\partial}_0\underline{h}_{00} + \underline{m}^{0b}\underline{\partial}_0\underline{h}_{0b}\big)\\
&= 4\big(\underline{m}^{a\alpha'}\underline{\partial}_a\underline{h}_{\alpha'0} + \underline{m}^{00}\underline{\partial}_0\underline{h}_{00}\big)\big(\underline{m}^{b\beta'}\underline{\partial}_b\underline{h}_{\beta'0} + \underline{m}^{00}\underline{\partial}_0\underline{h}_{00} + \underline{m}^{0b}\underline{\partial}_0\underline{h}_{0b}\big)\\
&\quad + 4\underline{m}^{0a'}\underline{\partial}_0\underline{h}_{0a'}\big(\underline{h}^{b\beta'}\underline{\partial}_b\underline{h}_{\beta'0} + \underline{m}^{00}\underline{\partial}_0\underline{h}_{00}\big) + 4\underline{m}^{0a'}\underline{\partial}_0\underline{h}_{0a'}\underline{m}^{0b}\underline{\partial}_0\underline{h}_{0b}.
\end{aligned}
$$

The last term is already presented in the (4.34). The remaining terms are null quadratic terms (recall that $\underline{m}^{00} = (s/t)^2$). □

Now we combine Lemma 4.4 with Lemmas 4.5 and 4.8.

Lemma 4.9. *There exists a positive constant $\varepsilon_w > 0$ such that if $|h|+|\partial h| \leq \varepsilon_w$, then the quasi-null term $\underline{P}_{00}$ is a linear combination of the following terms with smooth and homogeneous coefficients of order ≤ 0:*

$$GQS_h(0,0), \quad Cub(0,0), \quad Com(0,0), \quad \partial_t\underline{h}_{a\alpha}\partial_t\underline{h}_{b\beta}. \tag{4.37}$$

The term $\underline{P}_{a\beta}$ is a linear combination of the following terms with smooth and homogeneous coefficients of order ≤ 0:

$$GQS_h(0,0), \quad Cub(0,0), \quad Com(0,0). \tag{4.38}$$

Proof. In view of Lemma 4.4, we need to focus on $\underline{g}^{\gamma\gamma'}\underline{g}^{\delta\delta'}\partial_t\underline{h}_{\gamma\gamma'}\partial_t\underline{h}_{\delta\delta'}$ and $\underline{m}^{\gamma\gamma'}\underline{m}^{\delta\delta'}\partial_t\underline{h}_{\gamma\delta}\partial_t\underline{h}_{\gamma'\delta'}$. The first term is covered by Lemma 4.8 and the second term is bounded as follows: we recall that

$$\big|\partial^I L^J \underline{m}^{00}\big| = C(I,J)(s/t)^2, \quad \big|\underline{m}^{\alpha\beta}\big| \leq C.$$

Then, when $(\gamma,\gamma') = (0,0)$ or $(\delta,\delta') = (0,0)$, we have $\underline{m}^{\gamma\gamma'}\underline{m}^{\delta\delta'}\partial_t\underline{h}_{\gamma\delta}\partial_t\underline{h}_{\gamma'\delta'}$ becomes a null term. When $(\gamma,\gamma') \neq (0,0)$ and $(\delta,\delta') \neq (0,0)$, we denote by $(\gamma,\gamma') = (a,\alpha)$ and $(\delta,\delta') = (b,\beta)$, so we see that $\underline{m}^{\gamma\gamma'}\underline{m}^{\delta\delta'}\partial_t\underline{h}_{\gamma\delta}\partial_t\underline{h}_{\gamma'\delta'}$ is a linear combination of $\partial_t\underline{h}_{a\alpha}\partial_t\underline{h}_{b\beta}$ with homogeneous coefficients of degree zero. □

Finally, we emphasize that, in order to control the quasi-null terms, we must control the term $\partial_t \underline{h}_{a\alpha} \partial_t \underline{h}_{b\beta}$ which is *not* a null term. This term will be bounded by refined decay estimates on $\partial \underline{h}_{a\alpha}$, and we refer to our forthcoming analysis in Section 9.

Chapter 5

Initialization of the Bootstrap Argument

5.1 The bootstrap assumption and the basic estimates

The bootstrap assumption

From now on, we assume that in a hyperbolic time interval $[2, s^*]$, the following energy bounds hold for $|I| + |J| \leq N$. Here $N \geq 14$, (C_1, ε) is a pair of positive constants and $1/50 \leq \delta \leq 1/20$, say.

$$E_M^*(s, \partial^I L^J h_{\alpha\beta})^{1/2} \leq C_1 \varepsilon s^\delta, \tag{5.1a}$$

$$E_{M,c^2}(s, \partial^I L^J \phi)^{1/2} \leq C_1 \varepsilon s^{1/2+\delta}. \tag{5.1b}$$

For $|I| + |J| \leq N - 4$ we have (in which (5.2a) is repeated from (5.1a) for clarity in the presentation)

$$E_M^*(s, \partial^I L^J h_{\alpha\beta})^{1/2} \leq C_1 \varepsilon s^\delta, \tag{5.2a}$$

$$E_{M,c^2}(s, \partial^I L^J \phi)^{1/2} \leq C_1 \varepsilon s^\delta. \tag{5.2b}$$

In combination with Lemma 3.3, we see that the total energy of $h_{\alpha\beta}$ on the hyperboloid $\mathcal{H}_s$ is bounded by

$$E_M(s, \partial^I L^J h_{\alpha\beta}) \leq CC_1 \varepsilon s^\delta + Cm_S \leq 2C_1 \varepsilon s^\delta, \tag{5.3}$$

where we take $m_S \leq \varepsilon$. In the following discussion, except if specified otherwise, the letter C always represents a constant depending only on N. This constant may change at each occurrence.

Basic L^2 estimates of the first generation

These estimates come directly from the above energy bounds.

For $|I| + |J| \leq N$, we have

$$\|(s/t)\partial_\gamma \partial^I L^J h_{\alpha\beta}\|_{L_f^2(\mathcal{H}_s)} + \|\underline{\partial}_a \partial^I L^J h_{\alpha\beta}\|_{L_f^2(\mathcal{H}_s)} \leq CC_1 \varepsilon s^\delta, \tag{5.4a}$$

$$\|(s/t)\partial_\alpha \partial^I L^J \phi\|_{L^2_f(\mathcal{H}_s)} + \|\underline{\partial}_a \partial^I L^J \phi\|_{L^2_f(\mathcal{H}_s)} \leq CC_1 \varepsilon s^{1/2+\delta}, \tag{5.4b}$$

$$\|\partial^I L^J \phi\|_{L^2_f(\mathcal{H}_s)} \leq CC_1 \varepsilon s^{1/2+\delta}. \tag{5.4c}$$

For $|I| + |J| \leq N - 1$, we have

$$\|\partial_\alpha \partial^I L^J \phi\|_{L^2_f(\mathcal{H}_s)} \leq CC_1 \varepsilon s^{1/2+\delta}. \tag{5.5}$$

For $|I| + |J| \leq N - 4$, we have

$$\|(s/t)\partial_\alpha \partial^I L^J \phi\|_{L^2_f(\mathcal{H}_s)} + \|\underline{\partial}_a \partial^I L^J \phi\|_{L^2_f(\mathcal{H}_s)} \leq CC_1 \varepsilon s^\delta \tag{5.6}$$

and, for $|I| + |J| \leq N - 5$,

$$\|\partial_\alpha \partial^I L^J \phi\|_{L^2_f(\mathcal{H}_s)} \leq CC_1 \varepsilon s^\delta. \tag{5.7}$$

Basic L^2 estimates of the second generation

These estimates come from the above L^2 bounds of the first generation combined with the commutator estimates presented in Lemma 3.15. For $|I| + |J| \leq N$, we obtain

$$\|(s/t)\partial^I L^J \partial_\gamma h_{\alpha\beta}\|_{L^2_f(\mathcal{H}_s)} + \|\partial^I L^J \underline{\partial}_a h_{\alpha\beta}\|_{L^2_f(\mathcal{H}_s)} \leq CC_1 \varepsilon s^\delta, \tag{5.8a}$$

$$\|(s/t)\partial^I L^J \partial_\alpha \phi\|_{L^2_f(\mathcal{H}_s)} + \|\partial^I L^J \underline{\partial}_a \phi\|_{L^2_f(\mathcal{H}_s)} \leq CC_1 \varepsilon s^{1/2+\delta}, \tag{5.8b}$$

while for $|I| + |J| \leq N - 1$ (the second term in the left-hand side being bounded by (4.17))

$$\|\partial^I L^J \partial_\alpha \phi\|_{L^2_f(\mathcal{H}_s)} + \|t\partial^I L^J \underline{\partial}_a \phi\|_{L^2_f(\mathcal{H}_s)} \leq CC_1 \varepsilon s^{1/2+\delta}. \tag{5.9}$$

For $|I| + |J| \leq N - 4$, we have

$$\|(s/t)\partial^I L^J \partial_\alpha \phi\|_{L^2_f(\mathcal{H}_s)} + \|\partial^I L^J \underline{\partial}_a \phi\|_{L^2_f(\mathcal{H}_s)} \leq CC_1 \varepsilon s^\delta, \tag{5.10}$$

while, for $|I| + |J| \leq N - 5$, again from (4.17))

$$\|\partial^I L^J \partial_\alpha \phi\|_{L^2_f(\mathcal{H}_s)} + \|t\partial^I L^J \underline{\partial}_a \phi\|_{L^2_f(\mathcal{H}_s)} \leq CC_1 \varepsilon s^\delta. \tag{5.11}$$

Basic L^∞ estimates of the first generation

For $|I|+|J|\leq N-2$, we obtain

$$\sup_{\mathcal{H}_s^*}\big(t^{3/2}(s/t)\partial_\gamma\partial^I L^J h_{\alpha\beta}\big)+\sup_{\mathcal{H}_s^*}\big(t^{3/2}\underline{\partial}_a\partial^I L^J h_{\alpha\beta}\big)\leq CC_1\varepsilon s^\delta, \tag{5.12a}$$

$$\sup_{\mathcal{H}_s}\big(t^{3/2}(s/t)\partial_\alpha\partial^I L^J\phi\big)+\sup_{\mathcal{H}_s}\big(t^{3/2}\underline{\partial}_a\partial^I L^J\phi\big)\leq CC_1\varepsilon s^{1/2+\delta}, \tag{5.12b}$$

$$\sup_{\mathcal{H}_s}\big(t^{3/2}\partial^I L^J\phi\big)\leq CC_1\varepsilon s^{1/2+\delta}. \tag{5.12c}$$

For $|I|+|J|\leq N-3$, we have

$$\sup_{\mathcal{H}_s}\big(t^{3/2}\partial_\alpha\partial^I L^J\phi\big)+\sup_{\mathcal{H}_s}\big(t^{5/2}\underline{\partial}_a\partial^I L^J\phi\big)\leq CC_1\varepsilon s^{1/2+\delta}. \tag{5.13}$$

Here, the second term in the left-hand side is bounded by applying (4.17) once more. For $|I|+|J|\leq N-6$, we have

$$\sup_{\mathcal{H}_s}\big(t^{3/2}(s/t)\partial_\alpha\partial^I L^J\phi\big)+\sup_{\mathcal{H}_s}\big(t^{3/2}\underline{\partial}_a\partial^I L^J\phi\big)\leq CC_1\varepsilon s^\delta, \tag{5.14}$$

while, for $|I|+|J|\leq N-7$,

$$\sup_{\mathcal{H}_s}\big(t^{3/2}\partial_\alpha\partial^I L^J\phi\big)+\sup_{\mathcal{H}_s}\big(t^{5/2}\underline{\partial}_a\partial^I L^J\phi\big)\leq CC_1\varepsilon s^\delta. \tag{5.15}$$

Basic L^∞ estimates of the second generation

For $|I|+|J|\leq N-2$, we obtain

$$\sup_{\mathcal{H}_s^*}\big(t^{1/2}|\partial^I L^J\partial_\gamma h_{\alpha\beta}|\big)\leq CC_1\varepsilon s^{-1+\delta},\quad \sup_{\mathcal{H}_s^*}\big(t^{3/2}|\partial^I L^J\underline{\partial}_a h_{\alpha\beta}|\big)\leq CC_1\varepsilon s^\delta, \tag{5.16a}$$

$$\sup_{\mathcal{H}_s}\big(t^{1/2}|\partial^I L^J\partial_\alpha\phi|\big)\leq CC_1\varepsilon s^{-1/2+\delta},\quad \sup_{\mathcal{H}_s}\big(t^{3/2}|\partial^I L^J\underline{\partial}_a\phi|\big)\leq CC_1\varepsilon s^{1/2+\delta}, \tag{5.16b}$$

$$\sup_{\mathcal{H}_s}\big(t^{3/2}|\partial^I L^J\phi|\big)\leq CC_1\varepsilon s^{1/2+\delta}. \tag{5.16c}$$

For $|I|+|J|\leq N-3$, we have

$$\sup_{\mathcal{H}_s}\big(t^{3/2}|\partial^I L^J\partial_\alpha\phi|\big)+\sup_{\mathcal{H}_s}\big(t^{5/2}|\partial^I L^J\underline{\partial}_a\phi|\big)\leq CC_1\varepsilon s^{1/2+\delta}, \tag{5.17}$$

while, for $|I|+|J|\leq N-6$,

$$\sup_{\mathcal{H}_s}\big(t^{1/2}|\partial^I L^J\partial_\alpha\phi|\big)\leq CC_1\varepsilon s^{-1+\delta},\qquad \sup_{\mathcal{H}_s}\big(t^{3/2}|\partial^I L^J\underline{\partial}_a\phi|\big)\leq CC_1\varepsilon s^\delta, \tag{5.18a}$$

$$\sup_{\mathcal{H}_s} \left(t^{3/2}|\partial^I L^J \phi|\right) \leq CC_1\varepsilon s^\delta. \tag{5.18b}$$

For $|I| + |J| \leq N - 7$, we find

$$\sup_{\mathcal{H}_s} \left(t^{3/2}|\partial^I L^J \partial_\alpha \phi|\right) + \sup_{\mathcal{H}_s} \left(t^{5/2}|\partial^I L^J \underline{\partial}_a \phi|\right) \leq CC_1\varepsilon s^\delta. \tag{5.19}$$

By (4.18) and (4.19), the following bounds hold:

$$\|\partial^I L^J \underline{\partial}_a \partial_{\beta'} h_{\alpha\beta}\|_{L^2(\mathcal{H}_s^*)} + \|\partial^I L^J \partial_{\beta'} \underline{\partial}_a h_{\alpha\beta}\|_{L^2(\mathcal{H}_s^*)} \leq CC_1\varepsilon s^{-1+\delta}, \tag{5.20}$$

$$\sup_{\mathcal{H}_s^*} \left(t^{3/2}|\partial^I L^J \underline{\partial}_a \partial_{\beta'} h_{\alpha\beta}|\right) + \sup_{\mathcal{H}_s^*} \left(t^{3/2}|\partial^I L^J \partial_{\beta'} \underline{\partial}_a h_{\alpha\beta}|\right) \leq CC_1\varepsilon s^{-1+\delta}. \tag{5.21}$$

5.2 Estimates based on integration along radial rays

For $|I| + |J| \leq N - 2$,

$$|\partial^I L^J h_{\alpha\beta}(t,x)| \leq CC_1\varepsilon(s/t)t^{-1/2}s^\delta + Cm_S t^{-1} \leq CC_1\varepsilon(s/t)t^{-1/2}s^\delta. \tag{5.22}$$

This estimate is based on the following observation:

$$\begin{aligned}
\left|\partial_r \partial^I L^J h_{\alpha\beta}(t,x)\right| &\leq C\left|\partial_\gamma \partial^I L^J h_{\alpha\beta}(t,x)\right| \\
&\leq CC_1\varepsilon t^{-1/2}s^{-1+\delta} \simeq CC_1\varepsilon t^{-1+\delta/2}(t-r)^{-1/2+\delta/2}.
\end{aligned}$$

Then we integrate $\partial_r \partial^I L^J h_{\alpha\beta}$ along the radial rays $\{(t,\lambda x)|1 \leq \lambda \leq (t-1)/|x|\}$. We see when $\lambda = (t-1)/|x|$, $\partial_r \partial^I L^J h_{\alpha\beta}(t,\lambda x) \simeq Cm_S t^{-1}$ since $h_{\alpha\beta}$ coincides with the Schwarzschild metric and, by integration, (5.22) holds.

Chapter 6

Direct Control of Nonlinearities in the Einstein Equations

6.1 L^∞ estimates

With the above estimates, we are in a position to control the good nonlinear terms: GQQ_{hh}, $GQQ_{h\phi}$, GQS_h, QS_ϕ, Com, and Cub.

Lemma 6.1. *When the basic sup-norm estimates hold, the following sup-norm estimates are valid for $k \le N-2$:*

$$\begin{aligned} |GQS_h(N-2,k)| &\le C(C_1\varepsilon)^2 t^{-2} s^{-1+2\delta}, \\ |GQQ_{hh}(N-2,k)| &\le C(C_1\varepsilon)^2 t^{-3} s^{2\delta}, \end{aligned} \tag{6.1}$$

$$|QS_\phi(N-2,k)| \le C(C_1\varepsilon)^2 t^{-2} s^{-1/2+2\delta}, \tag{6.2}$$

$$|GQQ_{h\phi}(N-2,k)| \le C(C_1\varepsilon)^2 t^{-3} s^{2\delta}, \tag{6.3}$$

$$|Com(N-2,k)| \le C(C_1\varepsilon)^2 t^{-5/2} s^{-1+2\delta}, \tag{6.4}$$

$$|Cub(N-2,k)| \le C(C_1\varepsilon)^2 t^{-5/2} s^{3\delta}. \tag{6.5}$$

Proof. We directly substitute the basic L^∞ estimates, and we begin

$$|GQS_h(N-2,k)| \le |(s/t)^2 \partial_t h \partial_t h| + \sum_{\substack{I_1+I_2=I \\ J_1+J_2=J}} |\partial^{I_1} L^{J_1} \underline{\partial}_a h_{\alpha\beta} \partial^{I_2} L^{J_2} \underline{\partial}_\nu h_{\alpha'\beta'}|.$$

By the basic decay estimate (5.16a), $|GQS_h(N-2,k)|$ is bounded by $C(C_1\varepsilon)^2 t^{-2} s^{-1+2\delta}$. The estimate for GQQ_{hh} is similar, where (5.21) is applied, and we omit the details. The estimate for QS_ϕ is more delicate and we have $\partial^I L^J(\partial_\mu\phi\partial_\nu\phi) = \sum_{\substack{I_1+I_2=I \\ J_1+J_2=J}} \partial^{I_1} L^{J_1} \partial_\mu\phi \partial^{I_2} L^{J_2} \partial_\nu\phi$.

- $I_1 = I, J_1 = J$ then $|I_2| = |J_2| = 0 \le N-7$. Then we apply (5.16b) and (5.19) we have
$$|\partial^{I_1} L^{J_1} \partial_\mu\phi \partial^{I_2} L^{J_2} \partial_\nu\phi| \le C(C_1\varepsilon)^2 t^{-2} s^{-1/2+2\delta}.$$

- $N-3 \geq |I_1|+|J_1| \geq N-5$ then $|I_2|+|J_2| \leq 3 \leq N-6$, then we apply (5.17) and (5.18a).
- $|I_1|+|J_1| = N-6$, this leads us to $|I_2|+|J_2| \leq 4 \leq N-3$, then we apply (5.18a) and (5.17).
- $|I_1|+|J_1| \leq N-7$, this leads us to $|I_2|+|J_2| \leq N-2$, then we apply (5.19) and (5.16b).

The estimate of $\partial^I L^J(\phi^2)$ is similar and we omit the details.

The estimate for Com is much simpler, due to the additional decay t^{-1}. We apply the above estimates to QS_ϕ and the basic sup-norm estimate directly. For the cubic term, we will not analyze each type but point out that the worst higher-order term is $h_{\alpha\beta}(\partial\phi)^2$, since $\partial^I L^J \partial_\alpha \phi$ has a decay $\simeq t^{-3/2}s^{1/2+\delta}$, but this term is found to be bounded by $t^{-5/2}(s/t)s^{3\delta}$. □

6.2 L^2 estimates

Lemma 6.2. *One has*

$$\|GQQ_{hh}(N,k)\|_{L^2(\mathcal{H}^*_s)} \leq C(C_1\varepsilon)^2 s^{-3/2+2\delta}, \tag{6.6}$$

$$\|GQS_h(N,k)\|_{L^2(\mathcal{H}^*_s)} \leq C(C_1\varepsilon)^2 s^{-3/2+2\delta}, \tag{6.7}$$

$$\|QS_\phi(N-4,k)\|_{L^2(\mathcal{H}^*_s)} \leq C(C_1\varepsilon)^2 s^{-3/2+2\delta}, \tag{6.8}$$

$$\|GQQ_{h\phi}(N-4,k)\|_{L^2(\mathcal{H}^*_s)} \leq C(C_1\varepsilon)^2 s^{-3/2+2\delta}, \tag{6.9}$$

$$\|Cub\|_{L^2_f(\mathcal{H}_s)} \leq C(C_1\varepsilon)^2 s^{-3/2+3\delta}. \tag{6.10}$$

Proof. For the term GQQ_{hh}, we will only write the estimate of $\partial^{I_1}L^{J_1}h_{\alpha'\beta'}\partial^{I_2}L^{J_2}\underline{\partial}_a\underline{\partial}_\nu h_{\alpha\beta}$ in detail and, to this end, we distinguish between two main cases:

Case 1. $|I_1| \geq 1$. *Subcase 1.1*: When $|I_1|+|J_1| \leq N-2$, we obtain

$$\begin{aligned}
\big\|\partial^{I_1}L^{J_1}h_{\alpha'\beta'}\partial^{I_2}L^{J_2}\underline{\partial}_a\underline{\partial}_\nu h_{\alpha\beta}\big\|_{L^2(\mathcal{H}^*_s)} &\leq CC_1\varepsilon\big\|t^{-1/2}s^{-1+\delta}\partial^{I_2}L^{J_2}\underline{\partial}_a\underline{\partial}_\nu h_{\alpha\beta}\big\|_{L^2(\mathcal{H}^*_s)}\\
&\leq CC_1\varepsilon s^{-3/2+\delta}E^*_M(s,\partial^{I_2}L^{J_2}\underline{\partial}h)^{1/2}\\
&\leq C(C_1\varepsilon)^2 s^{-3/2+2\delta}.
\end{aligned}$$

Subcase 1.2: When $N \geq |I_1|+|J_2| \geq N-1$, we have $|I_2|+|J_2| \leq 1 \leq N-3$, then in view of (5.20)

$$\begin{aligned}
&\big\|\partial^{I_1}L^{J_1}h_{\alpha'\beta'}\partial^{I_2}L^{J_2}\underline{\partial}_a\underline{\partial}_\nu h_{\alpha\beta}\big\|_{L^2(\mathcal{H}^*_s)}\\
&\leq CC_1\varepsilon\big\|t^{-3/2}s^{-1+\delta}(t/s)\big|(s/t)\partial^{I_1}L^{J_1}h_{\alpha'\beta'}\big\|_{L^2(\mathcal{H}^*_s)}\\
&\leq CC_1\varepsilon s^{-3/2+\delta}\big\|(s/t)\partial^{I_1}L^{J_1}h_{\alpha'\beta'}\big\|_{L^2(\mathcal{H}^*_s)} \leq C(C_1\varepsilon)^2 s^{-3/2+2\delta}.
\end{aligned}$$

Case 2. $|I_1| = 0$. *Subcase 2.1*: When $|J_1| \leq N-2$, then in view of (5.20) we obtain

$$\begin{aligned}
&\big\|L^{J_1} h_{\alpha'\beta'} \partial^I L^{J_2} \underline{\partial}_a \underline{\partial}_\nu h_{\alpha\beta}\big\|_{L^2(\mathcal{H}_s^*)} \\
&\leq CC_1\varepsilon \big\|\big((s/t)t^{-1/2}s^\delta + t^{-1}\big)\partial^I L^{J_2} \underline{\partial}_a \underline{\partial}_\nu h_{\alpha'\beta'}\|_{L^2(\mathcal{H}_s^*)} \\
&\leq CC_1\varepsilon \big\|\big((s/t)t^{-1/2}s^\delta + t^{-1}\big)s^{-1}\, |s\partial^I L^{J_2} \underline{\partial}_a \underline{\partial}_\nu h_{\alpha'\beta'}|\|_{L^2(\mathcal{H}_s^*)} \\
&\leq CC_1\varepsilon s^{-3/2+\delta} E_M^*(s, \partial^I L^{J_2} \underline{\partial} h)^{1/2} \leq C(C_1\varepsilon)^2 s^{-3/2+2\delta}.
\end{aligned}$$

Subcase 2.2: When $N \geq |J_1| \geq N-1 \geq 1$, then we denote by $L^{J_1} = L_a L^{J_1'}$, we have $|I| + |J_2| \leq 1 \leq N-3$. Then in view of (5.21)

$$\begin{aligned}
&\big\|L^{J_1} h_{\alpha'\beta'} \partial^I L^{J_2} \underline{\partial}_a \underline{\partial}_\nu h_{\alpha\beta}\big\|_{L^2(\mathcal{H}_s^*)} \\
&\leq CC_1\varepsilon \big\|t^{-3/2}s^{-1+\delta} L_a L^{J_1'} h_{\alpha'\beta'}\big\|_{L^2(\mathcal{H}_s^*)} \\
&\leq CC_1\varepsilon \big\|t^{-1/2}s^{-1+\delta} \underline{\partial}_a L^{J_1'} h_{\alpha'\beta'}\big\|_{L^2(\mathcal{H}_s^*)} \\
&\leq C(C_1\varepsilon)^2 s^{-3/2+2\delta}.
\end{aligned}$$

The estimate on the term GQS_h is similar, and we omit the details. For the estimate for $QS_\phi(N-4,k)$, we will only write the proof for $\partial^I L^J(\partial_\alpha\phi\partial_\beta\phi)$. For $N \geq 9$, we have $\left[\frac{N-4}{2}\right] \leq N-7$. So, at least $|I_1| + |J_1| \leq N-7$ or $|I_2| + |J_2| \leq N-7$:

$$\begin{aligned}
\big\|\partial^{I_1} L^{J_1} \partial_\alpha\phi\, \partial^{I_2} L^{J_2}\phi\big\|_{L^2(\mathcal{H}_s^*)} &\leq CC_1\varepsilon \big\|t^{-3/2}s^\delta (t/s)\,(s/t)\partial^{I_2} L^{J_2}\phi\big\|_{L^2(\mathcal{H}_s^*)} \\
&\leq C(C_1\varepsilon)^2 s^{-3/2+2\delta}.
\end{aligned}$$

As far as $GQQ_{h\phi}(N-4,k)$ is concerned, we only treat

$$\partial^{I_1} L^{J_1} h_{\alpha'\beta'} \partial^{I_2} L^{J_2} \underline{\partial}_a \underline{\partial}_\mu \phi.$$

We observe that $|I_1| + |J_1| \leq N-4$ and by applying (5.22)

$$\begin{aligned}
&\big\|\partial^{I_1} L^{J_1} h_{\alpha'\beta'} \partial^{I_2} L^{J_2} \underline{\partial}_a \underline{\partial}_\mu \phi\big\|_{L^2(\mathcal{H}_s^*)} \\
&\leq \Big\|\big((s/t)t^{-1/2}s^\delta + t^{-1}\big)s^{-1}\,\big(s\partial^{I_2} L^{J_2} \underline{\partial}_a \underline{\partial}_\mu \phi\big)\Big\|_{L^2(\mathcal{H}_s^*)} \\
&\leq CC_1\varepsilon s^{-3/2+\delta} \big\|s\partial^{I_2} L^{J_2} \underline{\partial}_a \underline{\partial}_\mu \phi\big\|_{L^2(\mathcal{H}_s^*)} \\
&\leq CC_1\varepsilon s^{-3/2+\delta} E_{M,c^2}\big(s, \partial^{I_2} L^{J_2} L_a \underline{\partial}_\mu \phi\big)^{1/2} \leq C(C_1\varepsilon)^2 s^{-3/2+2\delta}.
\end{aligned}$$

The higher-order terms Cub are bounded as we did for the sup-norm: just observe that the worst term is again $h(\partial\phi)^2$ and can be bounded as stated. □

Lemma 6.3. *For $N \geq 7$, one has*

$$\|QS_\phi(N,k)\|_{L^2(\mathcal{H}_s^*)} \leq C(C_1\varepsilon)^2 s^{-1+2\delta}. \tag{6.11}$$

Proof. We discuss the following cases:

- $|I_1| + |J_1| = N$, $N - 7 \geq 0$. So, in view of (5.8b) and (5.19):

$$\|\partial^{I_1} L^{J_1} \partial_\gamma \phi \partial^{I_2} L^{J_2} \partial_{\gamma'} \phi\|_{L^2(\mathcal{H}_s^*)} \leq CC_1\varepsilon \left\| t^{-3/2} s^\delta (t/s)\,(s/t) \partial^{I_1} L^{J_1} \partial_\gamma \phi \right\|_{L^2(\mathcal{H}_s^*)}$$
$$\leq CC_1 \varepsilon s^{-3/2+\delta}\, CC_1 \varepsilon s^{1/2+\delta} \leq C(C_1\varepsilon)^2 s^{-1+2\delta}.$$

- $|I_1| + |J_1| = N - 1$, then $|I_2| + |J_2| = 1 \leq N - 6$. So, in view of (5.9) and (5.18a), we have

$$\|\partial^{I_1} L^{J_1} \partial_\gamma \phi \partial^{I_2} L^{J_2} \partial_{\gamma'} \phi\|_{L^2(\mathcal{H}_s^*)} \leq CC_1\varepsilon \left\| t^{-1/2} s^{-1+\delta}\, \partial^{I_1} L^{J_1} \partial_\gamma \phi \right\|_{L^2(\mathcal{H}_s^*)}$$
$$\leq CC_1 \varepsilon s^{-3/2+\delta}\, CC_1 \varepsilon s^{1/2+\delta} \leq C(C_1\varepsilon)^2 s^{-1+2\delta}.$$

- $|I_1| + |J_1| = N - 2$, then $|I_2| + |J_2| = 2 \leq N - 5$. So, in view of (5.16a) and (5.11), we have

$$\|\partial^{I_1} L^{J_1} \partial_\gamma \phi \partial^{I_2} L^{J_2} \partial_{\gamma'} \phi\|_{L^2(\mathcal{H}_s^*)} \leq CC_1\varepsilon \left\| t^{-1/2} s^{-1/2+\delta}\, \partial^{I_2} L^{J_2} \partial_\gamma \phi \right\|_{L^2(\mathcal{H}_s^*)}$$
$$\leq CC_1 \varepsilon s^{-1+\delta}\, CC_1 \varepsilon s^{\delta} \leq C(C_1\varepsilon)^2 s^{-1+2\delta}.$$

- $|I_1| + |J_1| = N - 3$, then $|I_2| + |J_2| = 3 \leq N - 4$. So, in view of (5.17) and (5.10), we have

$$\|\partial^{I_1} L^{J_1} \partial_\gamma \phi \partial^{I_2} L^{J_2} \partial_{\gamma'} \phi\|_{L^2(\mathcal{H}_s^*)} \leq CC_1\varepsilon \left\| t^{-3/2} s^{1/2+\delta} (t/s)\,(s/t) \partial^{I_2} L^{J_2} \partial_\gamma \phi \right\|_{L^2(\mathcal{H}_s^*)}$$
$$\leq CC_1 \varepsilon s^{-1+\delta}\, CC_1 \varepsilon s^{\delta} \leq C(C_1\varepsilon)^2 s^{-1+2\delta}.$$

- When $|I_1| + |J_1| \leq N - 4 \leq 3$, we exchange the role of I_1, I_2 and J_1, J_2, and apply the arguments above again.

□

Chapter 7

Direct Consequences of the Wave Gauge Condition

7.1 L^∞ estimates

We now use the wave coordinate estimates (4.31) and (4.32). Combined with Proposition 3.6, they provide us with rather precise L^2 estimates and L^∞ estimate on the gradient of the metric coefficient $\underline{h}^{00}$. In view of these estimates, we can say (as in [41]) that the quasi-linear terms QQ_{hh} and $QQ_{h\phi}$ are essentially null terms. In $\mathcal{K}$, the gradient of a function u can be written in the semi-hyperboloidal frame, that is $\partial_\alpha u = \Psi_\alpha^{\alpha'} \underline{\partial}_{\alpha'} u = \Psi_\alpha^0 \partial_t u + \Psi_\alpha^a \underline{\partial}_a u$. The coefficients Ψ_α^β are smooth and homogeneous of degree 0. And we observe that the derivatives $\underline{\partial}_a$ are "good" derivatives. So our task is *to get refined estimates on $\partial_t u$*, which is the main purpose of the next subsections. We begin with the L^∞ estimates, whose derivation is simpler than the derivation of the L^2 estimates.

Lemma 7.1. *Assume that the bootstrap assumption* (5.1) *holds with $C_1\varepsilon$ sufficiently small so that Lemma 4.7 holds, then the following estimates hold for $|I| + |J| \leq N - 2$:*

$$|\partial^I L^J \partial_\alpha \underline{h}^{00}| + |\partial_\alpha \partial^I L^J \underline{h}^{00}| \leq CC_1\varepsilon t^{-3/2} s^\delta, \tag{7.1}$$

$$|\partial^I L^J \underline{h}^{00}| \leq CC_1\varepsilon t^{-1/2} (s/t)^2 s^\delta + Cm_s t^{-1}. \tag{7.2}$$

Proof. We derive (7.1) by substituting the basic sup-norm estimates into (4.32). Then we integrate (7.1) along radial rays, as we did in Section 5.2 and we obtain (7.2). □

The following statements are direct consequences of the above sup-norm estimates and play an essential role in our analysis. Roughly speaking, these lemmas guarantee that the curved metric g is sufficiently close to the Minkowski metric, so that the energy estimates in Propositions 3.1 and

3.2 hold, as well as a sup-norm estimate for the Klein-Gordon equation (discussed in Appendix C).

Lemma 7.2. (Equivalence between the curved energy and flat energy functionals). *Under the bootstrap assumption with $C_1\varepsilon$ sufficiently small so that Lemma 4.6 holds, there exists a constant $\kappa > 1$ such that*

$$\begin{aligned} \kappa^{-2} E_M^*(s, \partial^I L^J h_{\alpha\beta}) \leq E_g^*(s, \partial^I L^J h_{\alpha\beta}) \leq \kappa^2 E_M^*(s, \partial^I L^J h_{\alpha\beta}), \\ \kappa^{-2} E_{M,c^2}(s, \partial^I L^J \phi) \leq E_{g,c^2}(s, \partial^I L^J \phi) \leq \kappa^2 E_{M,c^2}(s, \partial^I L^J \phi). \end{aligned} \tag{7.3}$$

Proof. We only show the first statement, since the proof of the second one is similar. From the identity

$$\begin{aligned}
&E_g^*(s,u) - E_M^*(s,u) \\
&= \int_{\mathcal{H}_s^*} \Big(- h^{00}|\partial_t u|^2 + h^{ab}\partial_a u \partial_b u + \sum_a \frac{2x^a}{t} h^{a\beta}\partial_\beta u \partial_t u \Big)\, dx \\
&= \int_{\mathcal{H}_s^*} \Big(h^{\alpha\beta}\partial_\alpha u \partial_\beta u + 2\sum_a \frac{x^a}{t} h^{a\beta}\partial_t u \partial_\beta u - 2h^{0\beta}\partial_t u \partial_\beta u \Big)\, dx \\
&= \int_{\mathcal{H}_s^*} \Big(\underline{h}^{\alpha\beta}\underline{\partial}_\alpha u \underline{\partial}_\beta u + \sum_a \frac{2x^a}{t} \underline{h}^{\alpha'\beta'}\Phi_{\alpha'}^a \Phi_{\beta'}^\beta \partial_t u \partial_\beta u - 2\underline{h}^{\alpha'\beta'}\Phi_{\alpha'}^0 \Phi_{\beta'}^\beta \partial_t u \partial_\beta u \Big)\, dx
\end{aligned}$$

and then

$$\begin{aligned}
&E_g^*(s,u) - E_M^*(s,u) \\
&= \int_{\mathcal{H}_s^*} \underline{h}^{\alpha\beta}\underline{\partial}_\alpha u \underline{\partial}_\beta u dx + \int_{\mathcal{H}_s^*} \Big(\frac{2x^a}{t}\underline{h}^{a0}|\partial_t u|^2 + \frac{2x^a}{t}\underline{h}^{ab}\partial_t u \underline{\partial}_b u \Big)\, dx \\
&\quad + \int_{\mathcal{H}_s^*} \Big(- 2\underline{h}^{00}|\partial_t u|^2 - 2\underline{h}^{0b}\partial_t u \underline{\partial}_b u - \frac{2x^a}{t}\underline{h}^{a0}|\partial_t u|^2 - \frac{2x^a}{t}\underline{h}^{ab}\partial_t u \underline{\partial}_b u \Big)\, dx \\
&= \int_{\mathcal{H}_s^*} \big(- \underline{h}^{00}|\partial_t u|^2 + \underline{h}^{ab}\underline{\partial}_a u \underline{\partial}_b u \big)\, dx \\
&= \int_{\mathcal{H}_s^*} \big(- (t/s)^2 \underline{h}^{00}|(s/t)\partial_t u|^2 + \underline{h}^{ab}\underline{\partial}_a u \underline{\partial}_b u \big)\, dx,
\end{aligned}$$

we obtain

$$|E_g^*(s,u) - E_M^*(s,u)| \leq C\Big(\|(t/s)^2 \underline{h}^{00}\|_{L^\infty(\mathcal{H}_s^*)} + \sum_{a,b} \|\underline{h}^{ab}\|_{L^\infty(\mathcal{H}_s^*)} \Big) E_M^*(s,u).$$

Then, recall that in view of (7.2), $|h| \leq CC_1\varepsilon(s/t)t^{-1/2}s^\delta + Cm_S t^{-1}$. When $C_1\varepsilon$ is sufficiently small, we have

$$|\underline{h}^{\alpha\beta}| \leq C \max_{\alpha,\beta} |h_{\alpha\beta}| \leq CC_1\varepsilon(s/t)t^{-1/2}s^\delta + Cm_S t^{-1}. \tag{7.4}$$

On the other hand, from (7.2), we obtain $|\underline{h}^{00}| \leq CC_1\varepsilon(s/t)^2t^{-1/2}s^\delta + Cm_S t^{-1}$, which implies

$$|(t/s)^2\underline{h}^{00}| \leq CC_1\varepsilon t^{-1/2}s^\delta + Cm_S. \tag{7.5}$$

Now, when $C_1\varepsilon$ is sufficiently small, (7.4) and (7.5) imply that $|E_g^*(s,u) - E_M^*(s,u)| \leq (1/2)E_M^*(s,u)$, which leads us to the desired result. □

Lemma 7.3 (Derivation of the uniform bound on $M_{\alpha\beta}$). *Under the energy assumption* (5.2), *the following estimate holds:*

$$M_{\alpha\beta}[\partial^I L^J h] \leq C(C_1\varepsilon)^2 s^{-3/2+2\delta}, \quad |I|+|J| \leq N, \tag{7.6}$$

and

$$M[\partial^I L^J \phi] \leq C(C_1\varepsilon)^2 s^{-3/2+2\delta}, \quad |I|+|J| \leq N-4, \tag{7.7a}$$

$$M[\partial^I L^J \phi] \leq C(C_1\varepsilon)^2 s^{-1+2\delta}, \quad |I|+|J| \leq N. \tag{7.7b}$$

Proof. We only provide the proof of the third inequality, since the other two are easier. Recall the definition of $M[\partial^I L^J\phi]$

$$\begin{aligned}
&\int_{\mathcal{H}_s} (s/t)\big|\partial_\mu g^{\mu\nu}\partial_\nu(\partial^I L^J\phi)\partial_t(\partial^I L^J\phi) - \frac{1}{2}\partial_t g^{\mu\nu}\partial_\mu(\partial^I L^J\phi)\partial_\nu(\partial^I L^J\phi)\big|\,dx \\
&\leq M[\partial^I L^J\phi](s)E_M(s,\partial^I L^J\phi)^{1/2}.
\end{aligned} \tag{7.8}$$

We perform the following calculation:

$$\begin{aligned}
&(s/t)\partial_\mu g^{\mu\nu}\partial_\nu\left(\partial^I L^J\phi\right)\partial_t\left(\partial^I L^J\phi\right) = (s/t)\partial_\mu h^{\mu\nu}\partial_\nu\left(\partial^I L^J\phi\right)\partial_t\left(\partial^I L^J\phi\right) \\
&= (s/t)\underline{\partial}_\mu\underline{h}^{\mu\nu}\underline{\partial}_\nu\left(\partial^I L^J\phi\right)\partial_t\left(\partial^I L^J\phi\right) \\
&\quad - (s/t)\partial_{\mu'}\left(\Psi_\mu^{\mu'}\Psi_\nu^{\nu'}\right)h^{\mu\nu}\partial_{\nu'}\left(\partial^I L^J\phi\right)\partial_t\left(\partial^I L^J\phi\right) \\
&= (s/t)\partial_t\underline{h}^{00}\partial_t\left(\partial^I L^J\phi\right)\partial_t\left(\partial^I L^J\phi\right) \\
&\quad + (s/t)\partial_t\underline{h}^{0a}\underline{\partial}_a\left(\partial^I L^J\phi\right)\partial_t\left(\partial^I L^J\phi\right) + (s/t)\underline{\partial}_b\underline{h}^{b0}\partial_t\left(\partial^I L^J\phi\right)\partial_t\left(\partial^I L^J\phi\right) \\
&\quad + (s/t)\underline{\partial}_a\underline{h}^{ab}\underline{\partial}_b\left(\partial^I L^J\phi\right)\partial_t\left(\partial^I L^J\phi\right) \\
&\quad - (s/t)\partial_{\mu'}\left(\Psi_\mu^{\mu'}\Psi_\nu^{\nu'}\right)h^{\mu\nu}\partial_{\nu'}\left(\partial^I L^J\phi\right)\partial_t\left(\partial^I L^J\phi\right)
\end{aligned} \tag{7.9}$$

and then we observe that

$$
\begin{aligned}
&\int_{\mathcal{H}_s} (s/t) \left|\partial_t \underline{h}^{00} \partial_t \left(\partial^I L^J \phi\right) \partial_t \left(\partial^I L^J \phi\right)\right| dx \\
&= \int_{\mathcal{H}_s} (t/s) \left|\partial_t \underline{h}^{00}\right| \left|(s/t) \partial_t \left(\partial^I L^J \phi\right)\right|^2 dx \\
&\leq C C_1 \varepsilon \int_{\mathcal{H}_s} (t/s) t^{-3/2} s^{\delta} \left|(s/t) \partial_t \left(\partial^I L^J \phi\right)\right|^2 dx \\
&\leq C C_1 \varepsilon s^{-3/2+\delta} E_M(s, \partial^I L^J \phi) \\
&\leq \begin{cases} C(C_1\varepsilon)^2 s^{-3/2+2\delta} E_M(s, \partial^I L^J \phi)^{1/2}, & |I|+|J| \leq N-4, \\ C(C_1\varepsilon)^2 s^{-1+2\delta} E_M(s, \partial^I L^J \phi)^{1/2}, & N-3 \leq |I|+|J| \leq N, \end{cases}
\end{aligned}
$$

where we have used (7.1), (5.1b) and (5.2b). The second, third, and fourth terms in the right-hand side of (7.9) are null terms, we observe that the second term is bounded as follows:

$$
\begin{aligned}
&\int_{\mathcal{H}_s} \left|(s/t) \partial_t \underline{h}^{0a} \underline{\partial}_a \left(\partial^I L^J \phi\right) \partial_t \left(\partial^I L^J \phi\right)\right| dx \\
&\leq \int_{\mathcal{H}_s} \left|\partial_t \underline{h}^{0a}\right| \left|\underline{\partial}_a \left(\partial^I L^J \phi\right) (s/t) \partial_t \left(\partial^I L^J \phi\right)\right| dx \\
&\leq C C_1 \varepsilon s^{-3/2+\delta} E_M(s, \partial^I L^J \phi) \\
&\leq \begin{cases} C(C_1\varepsilon)^2 s^{-3/2+2\delta} E_M(s, \partial^I L^J \phi)^{1/2}, & |I|+|J| \leq N-4, \\ C(C_1\varepsilon)^2 s^{-1+2\delta} E_M(s, \partial^I L^J \phi)^{1/2}, & N-3 \leq |I|+|J| \leq N. \end{cases}
\end{aligned}
$$

The third and fourth terms are bounded similarly and we omit the details.

The last term is bounded by applying the additional decay provided by $\partial_{\mu'} \left(\Psi_\mu^{\mu'} \Psi_\nu^{\nu'}\right)$. This term is bounded by t^{-1}. We have

$$
\begin{aligned}
&\int_{\mathcal{H}_s} \left|(s/t) \partial_{\mu'} \left(\Psi_\mu^{\mu'} \Psi_\nu^{\nu'}\right) h^{\mu\nu} \partial_{\nu'} \left(\partial^I L^J \phi\right) \partial_t \left(\partial^I L^J \phi\right)\right| dx \\
&\leq C C_1 \varepsilon \int_{\mathcal{H}_s} t^{-1} (t/s) |h^{\mu\nu}| \left|(s/t) \partial_{\nu'} \left(\partial^I L^J \phi\right) (s/t) \partial_t \left(\partial^I L^J \phi\right)\right| dx \\
&\leq C C_1 \varepsilon \int_{\mathcal{H}_s} s^{-1} \left(t^{-1} + t^{-1/2} (s/t) s^{\delta}\right) \left|(s/t) \partial_{\nu'} \left(\partial^I L^J \phi\right) (s/t) \partial_t \left(\partial^I L^J \phi\right)\right| dx \\
&\leq C C_1 \varepsilon s^{-3/2+\delta} E_M(s, \partial^I L^J \phi) \\
&\leq \begin{cases} C(C_1\varepsilon)^2 s^{-3/2+2\delta} E_M(s, \partial^I L^J \phi)^{1/2}, & |I|+|J| \leq N-4, \\ C(C_1\varepsilon)^2 s^{-1+2\delta} E_M(s, \partial^I L^J \phi)^{1/2}, & N-3 \leq |I|+|J| \leq N. \end{cases}
\end{aligned}
$$

We conclude that

$$\int_{\mathcal{H}_s} \left|(s/t)\partial_\mu g^{\mu\nu}\partial_\nu\left(\partial^I L^J\phi\right)\partial_t\left(\partial^I L^J\phi\right)\right| dx$$
$$\leq \begin{cases} C(C_1\varepsilon)^2 s^{-3/2+2\delta} E_M(s,\partial^I L^J\phi)^{1/2}, & |I|+|J|\leq N-4, \\ C(C_1\varepsilon)^2 s^{-1+2\delta} E_M(s,\partial^I L^J\phi)^{1/2}, & N-3\leq |I|+|J|\leq N. \end{cases}$$

The term $\partial_t g^{\mu\nu}\partial_\mu(\partial^I L^J\phi)\partial_\nu(\partial^I L^J\phi)$ is bounded similarly and we omit the details. □

Lemma 7.4. *Following the notation in Proposition 3.5. When the bootstrap assumption* (5.1) *holds, the following estimate holds:*

$$|h'_{t,x}(\lambda)| \leq CC_1\varepsilon(s/t)^{1/2}\lambda^{-3/2+\delta} + CC_1\varepsilon(s/t)^{-1}\lambda^{-2}. \tag{7.10}$$

Proof. Following the notation in Proposition 3.5, we have

$$h_{t,x}(\lambda) = \overline{h}^{00}\left(\frac{\lambda t}{s}, \frac{\lambda x}{s}\right).$$

Recalling that $\overline{h}^{00} = (t/s)^2\underline{h}^{00}$, we find

$$h_{t,x}(\lambda) = (t/s)^2\underline{h}^{00}\left(\frac{\lambda t}{s}, \frac{\lambda x}{s}\right),$$

which gives us

$$h'_{t,x}(\lambda) = (t/s)^3\underline{\partial}_\perp\underline{h}^{00}\left(\frac{\lambda t}{s}, \frac{\lambda x}{s}\right). \tag{7.11}$$

Here we recall also that $\underline{\partial}_\perp\underline{h}^{00} = \frac{s^2}{t^2}\partial_t\underline{h}^{00} + \frac{x^a}{t}\underline{\partial}_a\underline{h}^{00} = \frac{s^2}{t^2}\partial_t\underline{h}^{00} + \frac{x}{t^2}L_a\underline{h}^{00}$. We see that, in view of (7.1), $\left|(t/s)\partial_t\underline{h}^{00}\right| \leq CC_1\varepsilon(s/t)^{1/2}s^{-3/2+\delta}$ and, in view of (7.2),

$$\left|(t/s)^2 s^{-1}L_a\underline{h}^{00}\right| \leq CC_1\varepsilon(s/t)^{1/2}s^{-3/2+\delta} + Cm_S ts^{-3}.$$

By combining this result with (7.11), the desired conclusion is reached. □

7.2 L^2 estimates

We first establish an L^2 estimate on the gradient of $\partial^I L^J\underline{h}^{00}$.

Lemma 7.5. *Under the bootstrap assumptions* (5.1) *and* (5.2), *the following estimate holds:*

$$\left\|\partial^I L^J\partial_\alpha\underline{h}^{00}\right\|_{L^2(\mathcal{H}_s^*)} + \left\|\partial_\alpha\partial^I L^J\underline{h}^{00}\right\|_{L^2(\mathcal{H}_s^*)} \leq CC_1\varepsilon s^{2\delta}. \tag{7.12}$$

Proof. The estimate is immediate in view of (4.32). Namely, thanks to the basic L^2 estimates, we have

$$\|(s/t)^2\partial\partial^{I'}L^{J'}\underline{h}\|_{L^2(\mathcal{H}^*_s)} + \|\underline{\partial}\partial^{I'}L^{J'}\underline{h}\|_{L^2(\mathcal{H}^*_s)} \leq CC_1\varepsilon s^\delta.$$

By (3.39), we get

$$\|t^{-1}\partial^I L^J \underline{h}^{00}\|_{L^2(\mathcal{H}^*_s)} \leq C\sum_a \|\underline{\partial}_a\partial^I L^J\underline{h}^{00}\|_{L^2(\mathcal{H}^*_s)} + Cm_S s^{-1} \leq CC_1\varepsilon s^\delta. \tag{7.13}$$

Now, from (4.32), we need to control the term $|\partial^{I_1}L^{J_1}\underline{h}\partial\partial^{I_2}L^{J_2}\underline{h}|$. When $|I_1|+|J_1|\leq N-2$, we apply (5.22) and (5.4a):

$$\|\partial^{I_1}L^{J_1}\underline{h}\partial\partial^{I_2}L^{J_2}\underline{h}\|_{L^2(\mathcal{H}^*_s)} \leq CC_1\varepsilon s^\delta\|(s/t)t^{-1/2}\partial^{I_2}L^{J_2}\underline{h}\|_{L^2(\mathcal{H}^*_s)} \leq CC_1\varepsilon s^\delta.$$

When $N-1\leq |I_1|+|J_1|\leq N$, we see that $|I_2|+|J_2|\leq 1$. We have

$$\begin{aligned}\|\partial^{I_1}L^{J_1}\underline{h}\partial\partial^{I_2}L^{J_2}\underline{h}\|_{L^2(\mathcal{H}^*_s)} &\leq CC_1\varepsilon s^\delta\|t^{-1/2}s^{-1}\partial^{I_1}L^{J_1}\underline{h}\|_{L^2(\mathcal{H}^*_s)}\\ &\leq CC_1\varepsilon s^\delta\|t^{-1}\partial^{I_1}L^{J_1}\underline{h}\|_{L^2(\mathcal{H}^*_s)} \leq CC_1\varepsilon s^{2\delta},\end{aligned}$$

where we have used (7.13). □

We are going now to derive the L^2 estimate on (the "essential part" of) $\partial^I L^J\underline{h}^{00}$. This is one of the most challenging terms and we first decompose $\underline{h}^{00}$ as follows:

$$\underline{h}^{00} := \chi(r/t)\underline{h}_0^{00} + \underline{h}_1^{00},$$

where $\underline{h}_0^{00} = \underline{h}_S^{00}$ is the corresponding component of the Schwarzschild metric and the function χ is smooth with $\chi(\tau)=0$ for $\tau\in[0,1/3]$ while $\chi(\tau)=1$ for $\tau\geq 2/3$. We introduce the notation $\underline{h}_0^{00} := \chi(r/t)\underline{h}_S^{00}$ and an explicit calculation shows that in $\mathcal{K}_{[2,+\infty)}$

$$|\underline{h}_0^{00}| \leq Cm_S t^{-1} \leq Cm_S(1+r)^{-1}, \qquad |\partial_\alpha\underline{h}_0^{00}| \leq Cm_S t^{-2} \leq Cm_S(1+r^2)^{-1}.$$

This leads us to the estimate

$$\|\partial_a\underline{h}_0^{00}\|_{L^2_f(\mathcal{H}_s)} \leq Cm_S, \qquad \|\underline{\partial}_a\underline{h}_0^{00}\|_{L^2_f(\mathcal{H}_s)} \leq Cm_S \tag{7.14}$$

and we are ready to establish the following result.

Proposition 7.1. *Assume that the bootstrap assumptions* (5.1) *and* (5.2) *hold with $C_1\varepsilon$ sufficiently small (so that Lemma 4.7 holds). Then, one has*

$$|\partial^I L^J\underline{h}^{00}| \leq Cm_S t^{-1} + |\partial^I L^J\underline{h}_1^{00}| \tag{7.15}$$

and

$$\begin{aligned}&\|(s/t)^{-1+\delta}s^{-1}\partial^I L^J\underline{h}_1^{00}\|_{L^2(\mathcal{H}^*_s)}\\ &\leq CC_0\,\varepsilon + C\sum_{\substack{|I'|\leq|I|,|J'|\leq|J|\\ \alpha,\beta}} E^*_M(s,\partial^I L^J h_{\alpha\beta})^{1/2}\\ &\quad + C\sum_{\substack{|I'|\leq|I|,|J'|\leq|J|\\ \alpha,\beta}}\int_2^s \tau^{-1}E^*_M(\tau,\partial^{I'}L^{J'}h_{\alpha\beta})^{1/2}d\tau \leq CC_1\varepsilon s^\delta.\end{aligned} \tag{7.16}$$

Proof. In the decomposition of $\underline{h}^{00}$, the term $\partial_\alpha \partial^I L^J \underline{h}_1^{00}$ vanishes near the boundary of $\mathcal{K}_{[2,s^*]}$, since in a neighborhood of this boundary, $\underline{h}^{00} = \underline{h}_S^{00} = \underline{h}_0^{00}$. Furthermore, we have

$$\begin{aligned}&\|(s/t)^\delta \partial_\alpha \partial^I L^J \underline{h}_1^{00}\|_{L^2(\mathcal{H}_s^*)}\\&\leq \|(s/t)^\delta \partial_\alpha \partial^I L^J \underline{h}^{00}\|_{L^2(\mathcal{H}_s^*)} + \|(s/t)^\delta \partial_\alpha \partial^I L^J \underline{h}_0^{00}\|_{L^2(\mathcal{H}_s^*)}.\end{aligned} \tag{7.17}$$

We recall that $\partial_a = -\frac{x^a}{t}\partial_t + \underline{\partial}_a$, that is, ∂_α is a linear combination of ∂_t and $\underline{\partial}_a$ with homogeneous coefficients of degree 0, so the following estimates are direct in view of (4.32):

$$\begin{aligned}&\|(s/t)^\delta \partial_\alpha \partial^I L^J \underline{h}^{00}\|_{L^2(\mathcal{H}_s^*)}\\&\leq C \sum_{\substack{|I'|+|J'|\leq |I|+|J|\\|J'|\leq|J|}} \Big(\|(s/t)^2 \partial\partial^{I'} L^{J'} h\|_{L^2(\mathcal{H}_s^*)} + \|\underline{\partial}\partial^{I'} L^{J'} h\|_{L^2(\mathcal{H}_s^*)}\\&\qquad\qquad + \|t^{-1}\partial^{I'} L^{J'} h\|_{L^2(\mathcal{H}_s^*)}\Big)\\&\quad + C \sum_{\substack{|I_1|+|I_2|\leq|I|\\|J_1|+|J_2|\leq|J|}} \big\|(s/t)^\delta \partial^{I_1} L^{J_1} \underline{h}\partial\partial^{I_2} L^{J_2} \underline{h}\big\|_{L^2_f(\mathcal{H}_s)}.\end{aligned} \tag{7.18}$$

Here the first sum in the right-hand side is easily controlled by

$$\sum_{\substack{|I'|\leq|I|,|J'|\leq|J|\\\alpha,\beta}} E_M^*(s, \partial^{I'} L^{J'} h_{\alpha\beta})^{1/2} + C\|t^{-1}\partial^{I'} L^{J'} h\|_{L^2(\mathcal{H}_s^*)}.$$

For the last term, we observe that when $N \geq 3$, either $|I_1| + |J_1| \leq N-2$ or else $|I_2| + |J_2| \leq N-2$. When $|I_1| + |J_1| \leq N-2$, in view of (5.22),

$$\begin{aligned}&\big\|(s/t)^\delta \partial^{I_1} L^{J_1} \underline{h}\partial\partial^{I_2} L^{J_2} \underline{h}\big\|_{L^2_f(\mathcal{H}_s)}\\&\leq CC_1\varepsilon\big\|\big((s/t)t^{-1/2}s^\delta + t^{-1}\big)\partial^{I_2} L^{J_2} \partial\underline{h}\big\|_{L^2_f(\mathcal{H}_s)}\\&\leq CC_1\varepsilon\big\|(s/t)\partial^{I_2} L^{J_2}\partial\underline{h}\big\|_{L(\mathcal{H}_s)} \leq CC_1\varepsilon \sum_{\substack{|I'|\leq|I|,|J'|\leq|J|\\\alpha,\beta}} E_M^*(s, \partial^{I'} L^{J'} h_{\alpha\beta})^{1/2}.\end{aligned}$$

When $|I_2| + |J_2| \leq N-2$, we see that $|I_1| + |J_1| \geq 1$. Then we need to distinguish between two different cases. If $|I_1| \geq 1$, then

$$\begin{aligned}&\big\|(s/t)^\delta \partial^{I_1} L^{J_1} \underline{h}\partial\partial^{I_2} L^{J_2} \underline{h}\big\|_{L^2_f(\mathcal{H}_s)} \leq CC_1\varepsilon\big\|t^{-1/2}s^{-1+\delta}(s/t)^\delta \partial^{I_1} L^{J_1} \underline{h}\big\|_{L^2_f(\mathcal{H}_s)}\\&\leq CC_1\varepsilon\|t^{1/2}s^{-2+\delta}(s/t)^\delta (s/t)\partial^{I_1} L^{J_1} \underline{h}\|_{L^2(\mathcal{H}_s^*)}\\&\leq CC_1\varepsilon s^{-1} \sum_{\substack{|I'|\leq|I|,|J'|\leq|J|\\\alpha,\beta}} E_M^*(s, \partial^{I'} L^{J'} h_{\alpha\beta})^{1/2}.\end{aligned}$$

When $|I_1| = 0$, we see that $|J_1| \geq 1$. In this case we set $L^{J_1} = L_a L^{J_1'}$ with $|J_1'| \geq 1$. Then one has

$$\begin{aligned}
&\big\|(s/t)^\delta \partial^{I_1} L^{J_1} \underline{h} \partial \partial^{I_2} L^{J_2} \underline{h}\big\|_{L^2_f(\mathcal{H}_s)} \\
&\leq CC_1\varepsilon \big\|(s/t)^\delta t^{-1/2} s^{-1+\delta} L_a L^{J_1'} \underline{h}\big\|_{L^2_f(\mathcal{H}_s)} \\
&= CC_1\varepsilon \big\|(s/t)^\delta t^{-1/2} s^{-1+\delta} t \underline{\partial}_a L^{J_1'} \underline{h}\big\|_{L^2_f(\mathcal{H}_s)} \\
&= CC_1\varepsilon \big\|t^{1/2-\delta} s^{-1+2\delta} \underline{\partial}_a L^{J_1'} \underline{h}\big\|_{L^2_f(\mathcal{H}_s)} \\
&\leq CC_1\varepsilon \sum_{\substack{|I'|\leq|I|,|J'|\leq|J| \\ \alpha,\beta}} E^*_M(s, \partial^{I'} L^{J'} h_{\alpha\beta})^{1/2}.
\end{aligned}$$

Then the above discussion leads us to

$$\begin{aligned}
&\|(s/t)^\delta \partial_\alpha \partial^I L^J \underline{h}^{00}\|_{L^2(\mathcal{H}^*_s)} \\
&\leq \sum_{\substack{|I'|\leq|I|,|J'|\leq|J| \\ \alpha,\beta}} E^*_M(s, \partial^{I'} L^{J'} h_{\alpha\beta})^{1/2} + C\|t^{-1}\partial^{I'} L^{J'} h\|_{L^2(\mathcal{H}^*_s)}.
\end{aligned} \tag{7.19}$$

Now based on (7.19), we continue our discussion. Recalling the adapted Hardy inequality (3.39), we obtain

$$\|t^{-1}\partial^{I'} L^{J'} h\|_{L^2(\mathcal{H}^*_s)} \leq \|r^{-1}\partial^{I'} L^{J'} h\|_{L^2(\mathcal{H}^*_s)} \leq C\|\underline{\partial}\partial^I L^J h\|_{L^2(\mathcal{H}^*_s)} + Cm_S s^{-1},$$

so that

$$\|(s/t)^\delta \partial_\alpha \partial^I L^J \underline{h}^{00}\|_{L^2(\mathcal{H}^*_s)} \leq C \sum_{\substack{|I'|\leq|I|,|J'|\leq|J| \\ \alpha,\beta}} E^*_M(s, \partial^{I'} L^{J'} h_{\alpha\beta})^{1/2} + Cm_S s^{-1}.$$

On the other hand, by explicit calculation we have $\|\partial_\alpha \partial^I L^J \underline{h}_0^{00}\|_{L^2(\mathcal{H}^*_s)} \leq Cm_S s^{-1}$. So in view of (7.17)

$$\|(s/t)^\delta \partial_\alpha \partial^I L^J \underline{h}_1^{00}\|_{L^2(\mathcal{H}^*_s)} \leq C \sum_{\substack{|I'|\leq|I|,|J'|\leq|J| \\ \alpha,\beta}} E^*_M(s, \partial^{I'} L^{J'} h_{\alpha\beta})^{1/2} + Cm_S s^{-1}.$$

We also recall that by the basic L^2 estimate, $\|\underline{\partial}_a \partial^I L^J \underline{h}_1^{00}\|_{L^2_f(\mathcal{H}_s)} \leq CC_1\varepsilon s^\delta$. By Proposition 3.6 with $\sigma = 1 - \delta$, the desired result is established. □

7.3 Commutator estimates

Next, we use the basic estimates and the estimate for $\underline{h}^{00}$ in order to control the commutators $[\partial^I L^J, h^{\mu\nu}\partial_\mu\partial_\nu] h_{\alpha\beta}$.

Lemma 7.6. *Assume that the bootstrap assumptions* (5.1) *and* (5.2) *holds,* *then for $|I|+|J|\leq N-2$, the following estimate holds in $\mathcal{K}$:*

$$\begin{aligned}&\left|[\partial^I L^J, h^{\mu\nu}\partial_\mu\partial_\nu]h_{\alpha\beta}\right|\\ &\leq C(C_1\varepsilon)^2 t^{-2}s^{-1+2\delta} + CC_1\varepsilon\left(t^{-1}+(s/t)^2t^{-1/2}s^{\delta}\right)\sum_{|J'|<|J|}\left|\partial_t\partial_t\partial^I L^{J'}h_{\alpha\beta}\right|.\end{aligned} \tag{7.20}$$

Proof. We recall Lemma 4.3, to estimate $[\partial^I L^J, h^{\mu\nu}\partial_\mu\partial_\nu]h_{\alpha\beta}$, and we need to control the terms listed in (4.14). We see first that, in view of (6.1), $|GQQ_{hh}(p,k)|\leq C(C_1\varepsilon)^2t^{-3}s^{2\delta}$. For the term $t^{-1}\partial^{I_3}L^{J_3}h_{\mu\nu}\partial^{I_4}L^{J_4}\partial_\gamma h_{\mu'\nu'}$, we observe that $|I_3|+|I_4|\leq N-2$ and $|I_4|+|J_4|\leq N-2$, so

$$\begin{aligned}\left|t^{-1}\partial^{I_3}L^{J_3}h_{\mu\nu}\partial^{I_4}L^{J_4}\partial_\gamma h_{\mu'\nu'}\right| &\leq C(C_1\varepsilon)^2\left(t^{-1}+(s/t)t^{-1/2}s^{\delta}\right)t^{-1/2}s^{-1+\delta}\\ &\leq C(C_1\varepsilon)^2t^{-3}s^{2\delta}.\end{aligned}$$

For the term $\partial^{I_1}L^{J_1}\underline{h}^{00}\partial^{I_2}L^{J_2}\partial_t\partial_t h_{\alpha\beta}$, we see that $|I_1|+|J_1|\leq N-2$ and $|I_1|\geq 1$, $|I_2|+|J_2|\leq N-3$, so in view of (7.1)

$$\left|\partial^{I_1}L^{J_1}\underline{h}^{00}\partial^{I_2}L^{J_2}\partial_t\partial_t h_{\alpha\beta}\right|\leq CC_1\varepsilon s^{\delta}t^{-3/2}|\partial^{I_2}L^{J_2}\partial_t\partial_t h_{\alpha\beta}|. \tag{7.21}$$

For terms $L^{J_1'}\underline{h}^{00}\partial^I L^{J_2'}\partial_t\partial_t h_{\alpha\beta}$ and $\underline{h}^{00}\partial_\gamma\partial_{\gamma'}\partial^I L^{J'}h_{\alpha\beta}$, we first observe that by the condition $|J_2'|<|J|$ and $|J'|<|J|$, $|I|+|J_2'|\leq N-3$, $|I|+|J'|\leq N-3$. Then they are bounded by applying (7.2). We only write in detail $L^{J_1'}\underline{h}^{00}\partial^I L^{J_2'}\partial_t\partial_t h_{\alpha\beta}$:

$$\left|L^{J_1'}\underline{h}^{00}\partial^I L^{J_2'}\partial_t\partial_t h_{\alpha\beta}\right|\leq CC_1\varepsilon\left((s/t)^2t^{-1/2}s^{\delta}+t^{-1}\right)\sum_{|J'|<|J|}\left|\partial^I L^{J'}\partial_t\partial_t h_{\alpha\beta}\right|. \tag{7.22}$$

In view of the commutator estimate (3.54), we have $\left|\partial^I L^{J'}\partial_t\partial_t h_{\alpha\beta}\right|\leq C\sum_{\substack{\gamma,\gamma'\\|J''|\leq|J'|}}\left|\partial_\gamma\partial_{\gamma'}\partial^I L^{J''}h_{\alpha\beta}\right|$. We observe that (and this is an argument frequently applied in the following discussion, as it says that $\partial_t\partial_t$ is the only "bad" component of the Hessian):

$$\begin{aligned}\partial_t\partial_a u &= \partial_a\partial_t u = \underline{\partial}_a\partial_t u - \frac{x^a}{t}\partial_t\partial_t u,\\ \partial_a\partial_b u &= \underline{\partial}_a\underline{\partial}_b u - \frac{x^a}{t}\partial_t\underline{\partial}_b u - \frac{x^b}{t}\underline{\partial}_a\partial_t u + \frac{x^a x^b}{t^2}\partial_t\partial_t u\\ &\quad - \underline{\partial}_a\left(x^b/t\right)\partial_t u + \frac{x^a}{t}\partial_t\left(x^b/t\right)\partial_t u.\end{aligned} \tag{7.23}$$

Here we observe that the term $\partial_\gamma\partial_{\gamma'}\partial^I L^{J''}h_{\alpha\beta}$ is bounded by $\partial_t\partial_t\partial^I L^{J''}h_{\alpha\beta}$ plus other "good" terms. We see that, in $\mathcal{K}$, $\left|\partial_t\left(x^b/t\right)\right|\leq Ct^{-1}$, $\underline{\partial}_a\left(x^b/t\right)\leq Ct^{-1}$, so that

$$\left|\underline{\partial}_a\left(x^b/t\right)\partial_t\partial^I L^{J''}h_{\alpha\beta}\right| + \left|\frac{x^a}{t}\partial_t\left(x^b/t\right)\partial_t\partial^I L^{J''}h_{\alpha\beta}\right|\leq CC_1\varepsilon t^{-3/2}s^{-1+\delta}.$$

The terms $\underline{\partial}_a\partial_t\partial^I L^{J''}h_{\alpha\beta}$, $\partial_t\underline{\partial}_a\partial^I L^{J''}h_{\alpha\beta}$ and $\underline{\partial}_a\underline{\partial}_b\partial^I L^{J''}h_{\alpha\beta}$ are the second-order derivatives, where at least one derivative is "good" (i.e. $\underline{\partial}_a$). We apply (4.18), (4.19) and (4.20) and basic sup-norm estimate, then we conclude that these terms are bounded by $CC_1\varepsilon t^{-3/2}s^{-1+\delta}$. We conclude that

$$\left|\partial_\gamma\partial_{\gamma'}\partial^I L^{J''}h_{\alpha\beta}\right| \leq CC_1\varepsilon t^{-3/2}s^{-1+\delta} + \left|\partial_t\partial_t\partial^I L^{J''}h_{\alpha\beta}\right|. \tag{7.24}$$

Now we substitute this result into (7.22) and obtain

$$\begin{aligned}&\left|L^{J_1'}\underline{h}^{00}\partial^I L^{J_2'}\partial_t\partial_t h_{\alpha\beta}\right| \\ &\leq C(C_1\varepsilon)^2 t^{-3}s^{2\delta} + CC_1\varepsilon\left((s/t)^2 t^{-1/2}s^\delta + t^{-1}\right)\sum_{|J'|<|J|}\left|\partial_t\partial_t\partial^I L^{J'}\right|.\end{aligned}$$

By combining the estimates above, the desired result is proven. □

Lemma 7.7. *For $|I|+|J|\leq N$, one has*

$$\begin{aligned}&\left\|s[\partial^I L^J, h^{\mu\nu}\partial_\mu\partial_\nu]h_{\alpha\beta}\right\|_{L^2(\mathcal{H}^*_s)} \\ &\leq C(C_1\varepsilon)^2 s^{2\delta} + CC_1\varepsilon s^\delta\sum_{|J'|\leq 1}\left\|s^2(s/t)^{1-\delta}\partial^I L^{J'}\partial_t\partial_t h_{\alpha\beta}\right\|_{L^\infty(\mathcal{H}^*_s)} \\ &\quad + CC_1\varepsilon s^{1/2+\delta}\sum_{|J'|<|J|}\left\|(s/t)^{5/2}\partial_t\partial_t\partial^I L^{J'}h_{\alpha\beta}\right\|_{L^2(\mathcal{H}^*_s)}.\end{aligned} \tag{7.25}$$

Proof. The proof relies on Lemma 4.3 and we need to estimate the terms listed in (4.14). The term GQQ_{hh} is already bounded in view of (6.6). For the term $t^{-1}\partial^{I_1}L^{J_1}h_{\mu\nu}\partial^{I_2}L^{J_2}\partial_\gamma h_{\mu'\nu'}$, we have the following estimates. When $|I_1|+|J_1|\leq N-2$, we see that

$$\begin{aligned}&\left\|st^{-1}\partial^{I_1}L^{J_1}h_{\mu\nu}\partial^{I_2}L^{J_2}\partial_\gamma h_{\mu'\nu'}\right\|_{L^2_f(\mathcal{H}_s)} \\ &< \left\|\left(t^{-1}+t^{-1/2}(s/t)s^\delta\right)(s/t)\partial^{I_2}L^{J_2}\partial_\gamma h_{\mu'\nu'}\right\|_{L^2_f(\mathcal{H}_s)} \\ &\leq C(C_1\varepsilon)^2 s^{-1/2+2\delta}.\end{aligned}$$

When $|I_1|+|J_1|\geq N-1\geq 1$, we have $|I_2|+|J_2|\leq 1\leq N-2$. We distinguish between two subcases: when $|I_1|\geq 1$, we obtain

$$\begin{aligned}&\left\|st^{-1}\partial^{I_1}L^{J_1}h_{\mu\nu}\partial^{I_2}L^{J_2}\partial_\gamma h_{\mu'\nu'}\right\|_{L^2_f(\mathcal{H}_s)} \\ &\leq CC_1\varepsilon\left\|st^{-1}\partial^{I_1}L^{J_1}h_{\mu\nu}t^{-1/2}s^{-1+\delta}\right\|_{L^2_f(\mathcal{H}_s)} \leq C(C_1\varepsilon)^2 s^{-3/2+2\delta}.\end{aligned}$$

When $|I_1|=0$, then $|J_1|\geq 1$. We denote by $L^{J_1}=L_a L^{J_1'}$ and

$$\begin{aligned}&\left\|st^{-1}\partial^{I_1}L^{J_1}h_{\mu\nu}\partial^{I_2}L^{J_2}\partial_\gamma h_{\mu'\nu'}\right\|_{L^2_f(\mathcal{H}_s)} = \left\|s\underline{\partial}_a L^{J_1'}h_{\mu\nu}\partial^{I_2}L^{J_2}\partial_\gamma h_{\mu'\nu'}\right\|_{L^2_f(\mathcal{H}_s)} \\ &\leq CC_1\varepsilon\left\|s\underline{\partial}_a L^{J_1'}h_{\mu\nu}t^{-1/2}s^{-1+\delta}\right\|_{L^2_f(\mathcal{H}_s)} \leq C(C_1\varepsilon)^2 s^{-1/2+2\delta}.\end{aligned}$$

For the term $\partial^{I_1}L^{J_1}\underline{h}^{00}\partial^{I_2}L^{J_2}\partial_t\partial_t h_{\alpha\beta}$ with $|I_1|\geq 1$, we observe that

- When $1 \leq |I_1| + |J_1| \leq N-1$ we apply (7.1):

$$\begin{aligned}
&\left\|s\partial^{I_1}L^{J_1}\underline{h}^{00}\partial^{I_2}L^{J_2}\partial_t\partial_t h_{\alpha\beta}\right\|_{L^2(\mathcal{H}_s^*)}\\
&\leq CC_1\varepsilon\left\|st^{-3/2}s^{\delta}(t/s)\,(s/t)\partial^{I_2}L^{J_2}\partial_t\partial_t h_{\alpha\beta}\right\|_{L^2(\mathcal{H}_s^*)}\\
&\leq C(C_1\varepsilon)^2 s^{-1/2+2\delta}.
\end{aligned}$$

- When $|I_1| + |J_1| = N$, then $|I_2| + |J_2| = 0 \leq N-3$. So

$$\begin{aligned}
&\left\|s\partial^{I_1}L^{J_1}\underline{h}^{00}\partial^{I_2}L^{J_2}\partial_t\partial_t h_{\alpha\beta}\right\|_{L^2(\mathcal{H}_s^*)} \leq CC_1\varepsilon\left\|st^{-1/2}s^{-1+\delta}\,\partial^{I_1}L^{J_1}\underline{h}^{00}\right\|_{L^2(\mathcal{H}_s^*)}\\
&\leq CC_1\varepsilon s^{-1/2+\delta}\left\|\partial^{I_1}L^{J_1}\underline{h}^{00}\right\|_{L^2(\mathcal{H}_s^*)} \leq C(C_1\varepsilon)^2 s^{-1/2+3\delta},
\end{aligned}$$

where we have applied (7.12).

For the term $L^{J_1'}\underline{h}^{00}\partial^I L^{J_2'}\partial_t\partial_t h_{\alpha\beta}$, we apply the energy estimate to $L^J\underline{h}^{00}$ by Proposition 7.1 and the sup-norm estimate provided by Lemma 7.1.

- When $|J_1'| \leq N-2$, we apply (7.2)

$$\begin{aligned}
&\left\|sL^{J_1'}\underline{h}^{00}\partial^I L^{J_2'}\partial_t\partial_t h_{\alpha\beta}\right\|_{L^2(\mathcal{H}_s^*)}\\
&\leq CC_1\varepsilon\left\|s\left(t^{-1}+(s/t)^2t^{-1/2}s^{\delta}\right)\partial^I L^{J_2'}\partial_t\partial_t h_{\alpha\beta}\right\|_{L^2(\mathcal{H}_s^*)}\\
&\leq CC_1\varepsilon\left\|(s/t)\partial^I L^{J_2'}\partial_t\partial_t h_{\alpha\beta}\right\|_{L^2(\mathcal{H}_s^*)} + CC_1\varepsilon s^{1/2+\delta}\left\|(s/t)^{5/2}\partial^I L^{J_2'}\partial_t\partial_t h_{\alpha\beta}\right\|_{L^2(\mathcal{H}_s^*)}\\
&\leq C(C_1\varepsilon)^2 s^{\delta} + CC_1\varepsilon s^{1/2+\delta}\sum_{|J'|<|J|}\left\|(s/t)^{5/2}\partial^I L^{J'}\partial_t\partial_t h_{\alpha\beta}\right\|_{L^2(\mathcal{H}_s^*)}
\end{aligned}$$

- When $|J_1'| \geq N-1$, we apply Proposition 7.1

$$\begin{aligned}
&\left\|sL^{J_1'}\underline{h}^{00}\partial^I L^{J_2'}\partial_t\partial_t h_{\alpha\beta}\right\|_{L^2(\mathcal{H}_s^*)}\\
&\leq CC_1\varepsilon\left\|st^{-1}\partial^I L^{J_2'}\partial_t\partial_t h_{\alpha\beta}\right\|_{L^2(\mathcal{H}_s^*)} + \left\|sL^{J_1'}\underline{h}_1^{00}\partial^I L^{J_2'}\partial_t\partial_t h_{\alpha\beta}\right\|_{L^2(\mathcal{H}_s^*)}\\
&\leq C(C_1\varepsilon)^2 s^{\delta} + \left\|sL^{J_1'}\underline{h}_1^{00}\partial^I L^{J_2'}\partial_t\partial_t h_{\alpha\beta}\right\|_{L^2(\mathcal{H}_s^*)}\\
&\leq C(C_1\varepsilon)^2 s^{\delta} + \left\|(s/t)^{-1+\delta}s^{-1}L^{J_1'}\underline{h}_1^{00}\right\|_{L^2(\mathcal{H}_s^*)}\left\|s^2(s/t)^{1-\delta}\partial^I L^{J_2'}\partial_t\partial_t h_{\alpha\beta}\right\|_{L^\infty(\mathcal{H}_s^*)}\\
&\leq C(C_1\varepsilon)^2 s^{\delta} + CC_1\varepsilon s^{\delta}\sum_{|J'|\leq 1}\left\|s^2(s/t)^{1-\delta}\partial^I L^{J'}\partial_t\partial_t h_{\alpha\beta}\right\|_{L^\infty(\mathcal{H}_s^*)}.
\end{aligned}$$

For the term $\underline{h}^{00}\partial_\gamma\partial_{\gamma'}\partial^I L^{J'} h_{\alpha\beta}$, the estimate is similar. We apply (7.2) and

$$
\begin{aligned}
&\Big\|s\underline{h}^{00}\partial_\gamma\partial_{\gamma'}\partial^I L^{J'} h_{\alpha\beta}\Big\|_{L^2(\mathcal{H}^*_s)}\\
&\leq CC_1\varepsilon\Big\|(s/t)\partial_\gamma\partial_{\gamma'}\partial^I L^{J'} h_{\alpha\beta}\Big\|_{L^2(\mathcal{H}^*_s)} + \Big\|(s/t)^2 t^{-1/2} s^{1+\delta}\partial_\gamma\partial_{\gamma'}\partial^I L^{J'} h_{\alpha\beta}\Big\|_{L^2(\mathcal{H}^*_s)}\\
&\leq C(C_1\varepsilon)^2 s^\delta + CC_1\varepsilon s^{1/2+\delta}\sum_{|J'|<|J|}\Big\|(s/t)^{5/2}\partial_\gamma\partial_{\gamma'}\partial^I L^{J'} h_{\alpha\beta}\Big\|_{L^2(\mathcal{H}^*_s)}.
\end{aligned}
$$

Now we need to treat the last term and bound it by $\|(s/t)^{5/2}\partial_t\partial_t\partial^I L^{J'} h_{\alpha\beta}\|_{L^2(\mathcal{H}^*_s)}$. We rely on the discussion after (7.23) and conclude that

$$
\begin{aligned}
&\Big\|\underline{h}^{00}\partial_\gamma\partial_{\gamma'}\partial^I L^{J'} h_{\alpha\beta}\Big\|_{L^2(\mathcal{H}^*_s)}\\
&\leq \sum_{\substack{a,\mu\\ |J''|<|J'|}}\|\underline{h}^{00}\underline{\partial}_a\partial_\mu\partial^I L^{J''} h_{\alpha\beta}\|_{L^2(\mathcal{H}^*_s)} + C\sum_{|J''|<|J'|}\|\underline{h}^{00}\partial_t\partial_t\partial^I L^{J''} h_{\alpha\beta}\|_{L^2(\mathcal{H}^*_s)}\\
&\leq C(C_1\varepsilon)^2 s^{-1+\delta} + CC_1\varepsilon s^{-1/2+\delta}\sum_{|J''|<|J'|}\Big\|(s/t)^{5/2}\partial_t\partial_t\partial^I L^{J''} h_{\alpha\beta}\Big\|_{L^2(\mathcal{H}^*_s)}. \qquad \square
\end{aligned}
$$

Chapter 8

Second-Order Derivatives of the Spacetime Metric

8.1 Preliminary

We now establish L^2 and L^∞ bounds for the terms $\partial_t\partial_t\partial^I L^J h_{\alpha\beta}$ and $\partial^I L^J \partial_t\partial_t h_{\alpha\beta}$, which contain at least two partial derivatives ∂_t and which we refer informally to as "second-order derivatives". We can now apply the method in [39, Chapter 8]. However, we are here in a simpler situation, since the system is diagonalized with respect to second-order derivative terms. We recall the decomposition of the flat wave operator in the semi-hyperboloidal frame:

$$-\Box u = (s/t)^2\partial_t\partial_t u + 2\sum_a (x^a/t)\underline{\partial}_a\partial_t u - \sum_a \underline{\partial}_a\underline{\partial}_a u + \frac{r^2}{t^3}\partial_t u + \frac{3}{t}\partial_t u. \quad (8.1)$$

We also have the decomposition

$$h^{\mu\nu}\partial_\mu\partial_\nu h_{\alpha\beta} = \underline{h}^{\mu\nu}\underline{\partial}_\mu\underline{\partial}_\nu h_{\alpha\beta} + h^{\mu\nu}\partial_\mu\left(\Psi_\nu^{\nu'}\right)\underline{\partial}_{\nu'} h_{\alpha\beta}$$

of the curved part of the reduced wave operator. The main equation (1.6a) leads us to

$$\begin{aligned}
&\left((s/t)^2 - \underline{h}^{00}\right)\partial_t\partial_t h_{\alpha\beta} \\
&= -2\sum_a (x^a/t)\underline{\partial}_a\partial_t h_{\alpha\beta} + \sum_a \underline{\partial}_a\underline{\partial}_a h_{\alpha\beta} - \frac{r^2}{t^3}\partial_t h_{\alpha\beta} - \frac{3}{t}\partial_t h_{\alpha\beta} \\
&\quad + \underline{h}^{0a}\partial_t\underline{\partial}_a h_{\alpha\beta} + \underline{h}^{a0}\underline{\partial}_a\partial_t h_{\alpha\beta} + \underline{h}^{ab}\underline{\partial}_a\underline{\partial}_b h_{\alpha\beta} + h^{\mu\nu}\partial_\mu\left(\Psi_\nu^{\nu'}\right)\underline{\partial}_{\nu'} h_{\alpha\beta} \\
&\quad - F_{\alpha\beta} + 16\pi\partial_\alpha\phi\partial_\beta\phi + 8\pi c^2\phi^2 g_{\alpha\beta}.
\end{aligned} \quad (8.2)$$

Let us differentiate the equation (1.6a) with respect to $\partial^I L^J$, then by a similar procedure in the above discussion,

$$\begin{aligned}
&\left((s/t)^2-\underline{h}^{00}\right)\partial_t\partial_t\partial^I L^J h_{\alpha\beta}\\
&=-2\sum_a (x^a/t)\underline{\partial}_a\partial_t\partial^I L^J h_{\alpha\beta}+\sum_a \underline{\partial}_a\underline{\partial}_a\partial^I L^J h_{\alpha\beta}\\
&\quad-\frac{r^2}{t^3}\partial_t\partial^I L^J h_{\alpha\beta}-\frac{3}{t}\partial_t\partial^I L^J h_{\alpha\beta}\\
&\quad+\underline{h}^{0a}\partial_t\underline{\partial}_a\partial^I L^J h_{\alpha\beta}+\underline{h}^{a0}\underline{\partial}_a\partial_t\partial^I L^J h_{\alpha\beta}+\underline{h}^{ab}\underline{\partial}_a\underline{\partial}_b\partial^I L^J h_{\alpha\beta}\\
&\quad+h^{\mu\nu}\partial_\mu\left(\Psi_\nu^{\nu'}\right)\underline{\partial}_{\nu'}\partial^I L^J h_{\alpha\beta}\\
&\quad-\partial^I L^J F_{\alpha\beta}+[\partial^I L^J,h^{\mu\nu}\partial_\mu\partial_\nu]h_{\alpha\beta}+16\pi\partial^I L^J\left(\partial_\alpha\phi\partial_\beta\phi\right)\\
&\quad+8\pi c^2\partial^I L^J\left(\phi^2 g_{\alpha\beta}\right).
\end{aligned}\tag{8.3}$$

For convenience, we introduce the notation

$$\begin{aligned}
Sc_1[\partial^I L^J u]:=&-2\sum_a (x^a/t)\underline{\partial}_a\partial_t\partial^I L^J u+\sum_a \underline{\partial}_a\underline{\partial}_a\partial^I L^J u\\
&-\frac{r^2}{t^3}\partial_t\partial^I L^J u-\frac{3}{t}\partial_t\partial^I L^J u
\end{aligned}$$

and

$$\begin{aligned}
Sc_2[\partial^I L^J u]:=&\underline{h}^{0a}\partial_t\underline{\partial}_a\partial^I L^J u+\underline{h}^{a0}\underline{\partial}_a\partial_t\partial^I L^J u\\
&+\underline{h}^{ab}\underline{\partial}_a\underline{\partial}_b\partial^I L^J u+h^{\mu\nu}\partial_\mu\left(\Psi_\nu^{\nu'}\right)\underline{\partial}_{\nu'}\partial^I L^J u,
\end{aligned}$$

so that (8.2) becomes

$$\begin{aligned}
&\left((s/t)^2-\underline{h}^{00}\right)\partial_t\partial_t\partial^I L^J h_{\alpha\beta}=Sc_1[\partial^I L^J h_{\alpha\beta}]+Sc_2[\partial^I L^J h_{\alpha\beta}]\\
&-\partial^I L^J F_{\alpha\beta}+[\partial^I L^J,h^{\mu\nu}\partial_\mu\partial_\nu]h_{\alpha\beta}\\
&+16\pi\partial^I L^J\left(\partial_\alpha\phi\partial_\beta\phi\right)+8\pi c^2\partial^I L^J\left(\phi^2 g_{\alpha\beta}\right).
\end{aligned}\tag{8.4}$$

Now we apply the estimate (7.2) to $\underline{h}^{00}$ and see that when $t>2$ (which is the case if we are in $\mathcal{K}$) and $C_1\varepsilon$ sufficiently small, then

$$\begin{aligned}
(s/t)^2-\underline{h}^{00}&\geq (s/t)^2-CC_1\varepsilon\left((s/t)^2t^{-1/2}s^\delta+t^{-1}\right)\\
&=(s/t)^2\left(1-CC_1\varepsilon t^{-1/2}s^\delta-CC_1\varepsilon ts^{-2}\right)\geq\frac{1}{2}(s/t)^2.
\end{aligned}$$

This leads us to the following estimate. Later, this equation will be used to control the L^2 and L^∞ norms of $\partial_t\partial_t\partial^I L^J h_{\alpha\beta}$.

Lemma 8.1. *When $C_1\varepsilon$ is sufficiently small, the following estimate holds for all multi-indices (I,J):*

$$\begin{aligned}
\left|(s/t)^2\partial_t\partial_t\partial^I L^J h_{\alpha\beta}\right|\leq&\, C\left(\left|Sc_1[\partial^I L^J h_{\alpha\beta}]\right|+\left|Sc_2[\partial^I L^J h_{\alpha\beta}]\right|\right)\\
&+\left|\partial^I L^J F_{\alpha\beta}\right|+\left|QS_\phi(p,k)\right|\\
&+\left|[\partial^I L^J,h^{\mu\nu}\partial_\mu\partial_\nu]h_{\alpha\beta}\right|+|Cub(p,k)|.
\end{aligned}\tag{8.5}$$

8.2 L^∞ estimates

In this section, we apply (8.4) and the estimates of nonlinear terms presented in Lemma 6.1. First we need to establish the following pointwise estimates.

Lemma 8.2. *For any* (I, J), *the following pointwise estimate holds in* $\mathcal{K}$:

$$\left|Sc_1[\partial^I L^J u]\right| + \left|Sc_2[\partial^I L^J u]\right| \leq Ct^{-1} \sum_{|I'| \leq |I|, \alpha} \left|\partial_\alpha \partial^{I'} L^J u\right| + Ct^{-1} \sum_{a,\alpha} \left|\partial_\alpha \partial^I L_a L^J u\right|. \tag{8.6}$$

Proof. The estimate on the term Sc_1 is immediate by applying (4.18) and (4.19). The bound on Sc_2 is due to the fact that $\underline{h}^{\alpha\beta}$ are linear combinations of $h_{\alpha\beta}$ with smooth and homogeneous functions of degree zero plus higher-order corrections, which are bounded in $\mathcal{K}$. □

Lemma 8.3. *When the bootstrap assumption* (5.1) *and* (5.2) *hold, the following estimate holds in* $\mathcal{K}_{[2,s^*]}$:

$$|\partial_t \partial_t \partial^I L^J h_{\alpha\beta}| \leq CC_1 \varepsilon t^{1/2} s^{-3+2\delta}, \quad \textit{for} \quad |I| + |J| \leq N - 4. \tag{8.7}$$

Proof. The proof is a direct application of (8.5), where we neglect the higher-order term Cub. We just need to estimate each term in the right-hand side. We first observe that by the basic sup-norm estimate (5.12a) combined with (8.6)

$$\left|Sc_1[\partial^I L^J u]\right| + \left|Sc_2[\partial^I L^J u]\right| \leq CC_1 \varepsilon t^{-3/2} s^{-1+\delta}.$$

The estimate for $\partial^I L^J F_{\alpha\beta}$ can be expressed as $QS_h(p,k)$, $Cub(p,k)$, which is bounded by $|\partial^I L^J F_{\alpha\beta}| \leq C(C_1\varepsilon)^2 t^{-1} s^{-2+2\delta}$. The estimate on the commutator $[\partial^I L^J, h^{\mu\nu}\partial_\mu\partial_\nu] h_{\alpha\beta}$ is obtained by applying (7.20):

$$\begin{aligned} |[\partial^I L^J, h^{\mu\nu}\partial_\mu\partial_\nu] h_{\alpha\beta}| \leq & C(C_1\varepsilon)^2 t^{-2} s^{-1+2\delta} \\ & + CC_1\varepsilon \left(t^{-1} + (s/t)^2 t^{-1/2} s^\delta\right) \sum_{|J'|<|J|} \left|\partial_t \partial_t \partial^I L^{J'} h_{\alpha\beta}\right|. \end{aligned}$$

The estimate for QS_ϕ is derived as follows. We only estimate $\partial^I L^J\left(\partial_\alpha\phi\partial_\beta\phi\right)$, since dealing with the term $\partial^I L^J\left(\phi^2\right)$ is easier:

$$\left|\partial^I L^J \left(\partial_\alpha\phi\partial_\beta\phi\right)\right| \leq \sum_{\substack{|I_1|+|I_2|=I \\ |J_1|+|J_2|=J}} \left|\partial^{I_1} L^{J_1} \partial_\alpha\phi\, \partial^{I_2} L^{J_2} \partial_\beta\phi\right|.$$

Recalling that $|I| + |J| \leq N - 4$, we obtain:

- When $|I_1| + |J_1| \leq N - 7$,
$$|\partial^{I_1} L^{J_1} \partial_\alpha \phi \, \partial^{I_2} L^{J_2} \partial_\beta \phi| \leq CC_1\varepsilon |t^{-3/2} s^\delta| \, CC_1\varepsilon |t^{-1/2} s^{-1/2+\delta}|$$
$$\leq C(C_1\varepsilon)^2 t^{-2} s^{-1/2+2\delta}.$$
- When $N-6 \leq |I_1|+|J_1| \leq N-4$, we see that $|I_2|+|J_2| \leq 2 \leq N-7$ and
$$|\partial^{I_1} L^{J_1} \partial_\alpha \phi \, \partial^{I_2} L^{J_2} \partial_\beta \phi| \leq CC_1\varepsilon |t^{-1/2} s^{-1/2+\delta}| \, CC_1\varepsilon |t^{-3/2} s^\delta|$$
$$\leq C(C_1\varepsilon)^2 t^{-2} s^{-1/2+2\delta}.$$

So, we conclude that
$$|QS_\phi(N-4,k)| \leq C(C_1\varepsilon)^2 (s/t)^2 s^{-5/2+2\delta}.$$
We thus have
$$\begin{aligned}
&|(s/t)^2 \partial_t \partial_t \partial^I L^J h_{\alpha\beta}| \\
&\leq CC_1\varepsilon t^{-3/2} s^{-1+\delta} + C(C_1\varepsilon)^2 (s/t)^2 s^{-5/2+2\delta} \\
&\quad + CC_1\varepsilon \left(t^{-1} + (s/t)^2 t^{-1/2} s^\delta\right) \sum_{|J'|<|J|} |\partial_t \partial_t \partial^I L^{J'} h_{\alpha\beta}|.
\end{aligned} \tag{8.8}$$
Observe that when $|J| = 0$, the last term in the above estimate disappears and we conclude with (8.7). We proceed by induction on $|J|$. Assume that (8.7) holds for all $|J| \leq m - 1 < N - 4$. We will prove that it still holds for $|J| = m \leq N - 4$. We substitute (8.7) (case $|J'| < |J| = m$) into the last term of (8.8). □

8.3 L^2 estimates

The following two estimates are direct in view of (4.18) and (4.19) combined with the expression of the energy E_M^*.

Lemma 8.4. *For all multi-indices (I, J), one has*
$$\begin{aligned}
&\|\underline{\partial}_a \partial_\alpha \partial^I L^J h_{\alpha\beta}\|_{L^2(\mathcal{H}_s^*)} + \|\partial_\alpha \underline{\partial}_a \partial^I L^J h_{\alpha\beta}\|_{L^2(\mathcal{H}_s^*)} \\
&\leq Cs^{-1} E_M^*(s, \partial^I L_a L^J h_{\alpha\beta})^{1/2} + Cs^{-1} \sum_{|I'|\leq|I|,\gamma} E_M^*(s, \partial^{I'} L^J h_{\alpha\beta})^{1/2}.
\end{aligned} \tag{8.9}$$

A direct consequence of these bounds is that, for any (I, J),
$$\begin{aligned}
\|Sc_1[\partial^I L^J h_{\alpha\beta}]\|_{L^2(\mathcal{H}_s^*)} &\leq Cs^{-1} \sum_a E_M^*(s, \partial^I L_a L^J h_{\alpha\beta})^{1/2} \\
&\quad + Cs^{-1} \sum_{|I'|\leq|I|} E_M^*(s, \partial^{I'} L^J h_{\alpha\beta})^{1/2}.
\end{aligned} \tag{8.10}$$

This estimate will play an essential role in our forthcoming analysis. Our next task is the derivation of an L^2 estimate for Sc_2. The term $h^{\mu\nu}\partial_\mu\Psi_\nu^{\nu'}\underline{\partial}_{\nu'}h_{\alpha\beta}$ is bounded by the additional decay of $\left|\partial_\mu\Psi_\nu^{\nu'}\right| \leq t^{-1}$, and we thus focus on the first three quadratic terms. We provide the derive for the first term (but omit the second and third terms):

$$\begin{aligned}
&\left\|(t/s)^{3/2}h^{0a}\partial_t\underline{\partial}_a\partial^I L^J h_{\alpha\beta}\right\|_{L^2(\mathcal{H}_s^*)}\\
&\leq CC_1\varepsilon\left\|(t/s)^{3/2}\left(t^{-1}+(s/t)t^{-1/2}s^\delta\right)\partial_t\underline{\partial}_a\partial^I L^J h_{\alpha\beta}\right\|_{L^2(\mathcal{H}_s^*)}\\
&\leq CC_1\varepsilon s^{-1/2}\left\|\partial_t\underline{\partial}_a\partial^I L^J h_{\alpha\beta}\right\|_{L^2(\mathcal{H}_s^*)}+CC_1\varepsilon\left\|s^{-1/2+\delta}\partial_t\underline{\partial}_a\partial^I L^J h_{\alpha\beta}\right\|_{L^2(\mathcal{H}_s^*)}\\
&\leq CC_1\varepsilon s^{-1/2+\delta}\left\|\partial_t\underline{\partial}_a\partial^I L^J h_{\alpha\beta}\right\|_{L^2(\mathcal{H}_s^*)}.
\end{aligned}$$

Then we apply (8.9) and obtain

$$\begin{aligned}
&\left\|(t/s)^{3/2}h^{0a}\partial_t\underline{\partial}_a\partial^I L^J h_{\alpha\beta}\right\|_{L^2(\mathcal{H}_s^*)}\\
&\leq CC_1\varepsilon s^{-3/2+\delta}\sum_a E_M^*(s,\partial^I L_a L^J h_{\alpha\beta})^{1/2}\\
&\quad + CC_1\varepsilon s^{-3/2+\delta}\sum_{|I'|\leq|I|,\gamma} E_M^*(s,\partial^{I'} L^J h_{\alpha\beta})^{1/2}.
\end{aligned} \tag{8.11}$$

We conclude that

$$\begin{aligned}
&\left\|(t/s)^{3/2}Sc_2[\partial^I L^J h_{\alpha\beta}]\right\|_{L^2(\mathcal{H}_s^*)}\\
&\leq CC_1\varepsilon s^{-3/2+\delta}\sum_a E_M^*(s,\partial^I L_a L^J h_{\alpha\beta})^{1/2}\\
&\quad + CC_1\varepsilon s^{-3/2+\delta}\sum_{|I'|\leq|I|,\gamma} E_M^*(s,\partial^{I'} L^J h_{\alpha\beta})^{1/2}.
\end{aligned} \tag{8.12}$$

With the above preparation, in the rest of this subsection we will prove the following.

Lemma 8.5. *Under the bootstrap assumption* (5.1) *and* (5.2)

$$\|s^3t^{-2}\partial_t\partial_t\partial^I L^J h_{\alpha\beta}\|_{L^2(\mathcal{H}_s^*)} \leq CC_1\varepsilon s^{2\delta}, \qquad |I|+|J|\leq N-1. \tag{8.13}$$

Proof. *Step I. Estimates for the nonlinear terms.* The estimate of (8.13) is also based on Lemma 8.1.

1. This is done by direct application of (8.10) combined with the energy assumption:

$$\left\|Sc_1[\partial^I L^J h_{\alpha\beta}]\right\|_{L^2(\mathcal{H}_s^*)} \leq CC_1\varepsilon s^{-1+\delta}.$$

2. For the term Sc_2 is bounded in view of (8.12) combined with the energy assumption:

$$\big\|Sc_1[\partial^I L^J h_{\alpha\beta}]\big\|_{L^2(\mathcal{H}_s^*)} \leq C(C_1\varepsilon)^2 s^{-3/2+2\delta}.$$

3. Now we are about to estimate $\partial^I L^J F_{\alpha\beta}$. We observe that this term is a linear combination of $QS_h(p,k)$ and $Cub(p,k)$. We see that the term $QS_h(p,k)$ is bounded as follows:

$$\|QS_h(p,k)\|_{L^2(\mathcal{H}_s^*)} \leq \sum_{\substack{\alpha,\beta,\alpha'\beta'\\ \gamma,\gamma'}} \sum_{\substack{|I_1|+|I_2|\leq|I|\\ |J_1|+|J_2|\leq|J|}} \|\partial^{I_1}L^{J_1}\partial_\gamma h_{\alpha\beta}\,\partial^{I_2}L^{L_2}\partial_{\gamma'}h_{\alpha'\beta'}\|_{L^2(\mathcal{H}_s^*)}$$

When $N\geq 3$, we must have either $|I_1|+|J_1|\leq N-2$ or $|I_2|+|J_2|\leq N-2$. So

$$\begin{aligned}
&a\left\|\partial^{I_1}L^{J_1}\partial_\gamma h_{\alpha\beta}\,\partial^{I_2}L^{L_2}\partial_{\gamma'}h_{\alpha'\beta'}\right\|_{L^2(\mathcal{H}_s^*)}\\
&\leq CC_1\varepsilon\left\|t^{-1/2}s^{-1+\delta}\partial^{I_2}L^{L_2}\partial_{\gamma'}h_{\alpha'\beta'}\right\|_{L^2(\mathcal{H}_s^*)}\\
&\leq CC_1\varepsilon s^{\delta}\left\|(t/s)t^{-1/2}s^{-1+\delta}\,(s/t)\partial^{I_2}L^{L_2}\partial_{\gamma'}h_{\alpha'\beta'}\right\|_{L^2(\mathcal{H}_s^*)}\\
&\leq CC_1\varepsilon s^{-1+\delta}E_M^*(s,\partial^{I_2}L^{J_2}h_{\alpha'\beta'})^{1/2}\leq C(C_1\varepsilon)^2 s^{-1+2\delta}.
\end{aligned}$$

We can conclude that $\left\|\partial^I L^J F_{\alpha\beta}\right\|_{L^2(\mathcal{H}_s^*)} \leq C(C_1\varepsilon)^2 s^{-1+2\delta}$.

4. QS_ϕ is bounded directly in view of (6.10).

5. The estimate on the commutator is the most difficult. We combine the sup-norm estimate (8.7) and the estimate (7.25):

$$\begin{aligned}
&\left\|s[\partial^I L^J, h^{\mu\nu}\partial_\mu\partial_\nu]h_{\alpha\beta}\right\|_{L^2(\mathcal{H}_s^*)}\\
&\leq C(C_1\varepsilon)^2 s^{2\delta} + CC_1\varepsilon s^{\delta}\sum_{|J'|\leq 1}\left\|s^2(s/t)^{1-\delta}\partial^I L^{J_2'}\partial_t\partial_t h_{\alpha\beta}\right\|_{L^\infty(\mathcal{H}_s^*)}\\
&\quad + CC_1\varepsilon s^{1/2+\delta}\sum_{|J'|<|J|}\left\|(s/t)^{5/2}\partial_t\partial_t\partial^I L^{I'}h_{\alpha\beta}\right\|_{L^2(\mathcal{H}_s^*)}\\
&\leq C(C_1\varepsilon)^2 s^{2\delta} + C(C_1\varepsilon)^2 s^{\delta}\|s^2(s/t)^{1-\delta}t^{1/2}s^{-3+2\delta}\|_{L^\infty(\mathcal{H}_s^*)}\\
&\quad + CC_1\varepsilon s^{1/2+\delta}\sum_{|J'|<|J|}\left\|(s/t)^{5/2}\partial_t\partial_t\partial^I L^{J'}h_{\alpha\beta}\right\|_{L^2(\mathcal{H}_s^*)}\\
&\leq C(CC_1\varepsilon)^2 s^{2\delta} + CC_1\varepsilon s^{1/2+\delta}\sum_{|J'|<|J|}\left\|(s/t)^{5/2}\partial_t\partial_t\partial^I L^{J'}h_{\alpha\beta}\right\|_{L^2(\mathcal{H}_s^*)}.
\end{aligned}$$

We thus conclude Step 1 with the inequality

$$\begin{aligned}
&\left\|s^3t^{-2}\partial^I L^J\partial_t\partial_t h_{\alpha\beta}\right\|_{L^2(\mathcal{H}_s^*)}\\
&\leq CC_1\varepsilon s^{2\delta} + CC_1\varepsilon s^{1/2+\delta}\sum_{|J'|<|J|}\left\|(s/t)^{5/2}\partial_t\partial_t\partial^I L^{J'}h_{\alpha\beta}\right\|_{L^2(\mathcal{H}_s^*)}
\end{aligned}\tag{8.14}$$

and we remark that when $|J| = 0$ the last sum is empty.

Step II. Induction argument. For $|I| + |J| \leq N - 1$, we proceed by induction on $|J|$. When $|J| = 0$, the last term in (8.14) does not exist. Then in view of (8.5), we have

$$\left\|s^3 t^{-2} \partial_t \partial_t \partial^I L^J h_{\alpha\beta}\right\|_{L^2(\mathcal{H}_s^*)} \leq C(C_1\varepsilon) s^{2\delta}.$$

Then we assume that (8.13) holds for $|J| \leq n < N - 1$, we want to prove that it still holds for $|J| = n$. In this case, by our induction assumption, we have

$$\begin{aligned}
&\left\|s^3 t^{-2} \partial^I L^J \partial_t \partial_t h_{\alpha\beta}\right\|_{L^2(\mathcal{H}_s^*)} \\
&\leq C(CC_1\varepsilon)^2 s^{2\delta} + CC_1\varepsilon s^{1/2+\delta} \sum_{|J'|<|J|} \left\|(s/t)^{5/2} \partial_t \partial_t \partial^I L^{J'} h_{\alpha\beta}\right\|_{L^2(\mathcal{H}_s^*)} \\
&\leq C(C_1\varepsilon)^2 s^{2\delta}.
\end{aligned}$$

Then in view of (8.5), the desired result is established. □

8.4 Conclusion for general second-order derivatives

In the above subsection we have only estimate the terms of the form $\partial_t \partial_t \partial^I L^J h_{\alpha\beta}$, but we observe that by the identities (7.23) (and a similar argument below it in the proof of (7.7)) and the commutator estimates (3.54)

$$|\partial_\alpha \partial_\beta \partial^I L^J h_{\alpha\beta}| \leq CC_1\varepsilon t^{1/2} s^{-3+2\delta}, \qquad |I| + |J| \leq N - 4, \tag{8.15}$$

$$\|s^3 t^{-2} \partial_\alpha \partial_\beta \partial^I L^J h_{\alpha\beta}\|_{L^2(\mathcal{H}_s^*)} \leq CC_1\varepsilon s^{2\delta}, \qquad |I| + |J| \leq N - 1, \tag{8.16}$$

$$|\partial^I L^J \partial_\alpha \partial_\beta h_{\alpha\beta}| \leq CC_1\varepsilon t^{1/2} s^{-3+2\delta}, \qquad |I| + |J| \leq N - 4, \tag{8.17}$$

$$\|s^3 t^{-2} \partial^I L^J \partial_\alpha \partial_\beta h_{\alpha\beta}\|_{L^2(\mathcal{H}_s^*)} \leq CC_1\varepsilon s^{2\delta}, \qquad |I| + |J| \leq N - 1. \tag{8.18}$$

8.5 Commutator estimates

In this section, we improve the sup-norm and L^2 estimates for the commutators: our strategy is to apply Lemma 4.3.

Lemma 8.6. *Assume that the energy assumptions* (5.1) *and* (5.2) *hold, then for all* $|I| + |J| \leq N - 4$

$$\left|[\partial^I L^J, h^{\mu\nu} \partial_\mu \partial_\nu] h_{\alpha\beta}\right| \leq C(C_1\varepsilon)^2 t^{-2} s^{-1+3\delta} + C(C_1\varepsilon)^2 t^{-1/2} s^{-3+2\delta}, \tag{8.19}$$

while for all $|I|+|J|\leq N$

$$\begin{aligned}&\|s[\partial^I L^J, h^{\mu\nu}\partial_\mu\partial_\nu]h_{\alpha\beta}\|_{L^2(\mathcal{H}^*_s)}\\&\leq C(C_1\varepsilon)^2 s^{-1/2+3\delta} + CC_1\varepsilon\sum_{|J'|<|J|}\left\|s^3t^{-2}\partial_t\partial_t\partial^I L^{J'}h_{\alpha\beta}\right\|_{L^2(\mathcal{H}^*_s)}.\end{aligned}\tag{8.20}$$

Proof. The proof of (8.19) is immediate by combining (8.15) with (7.20). The proof of (8.20) relies on a refinement of the proof of (7.25). We will improve upon our estimates on $L^{J_1'}\underline{h}^{00}\partial^I L^{J_2'}\partial_t\partial_t h_{\alpha\beta}$ and $\underline{h}^{00}\partial^I L^{J'}h_{\alpha\beta}$. First we observe that for $L^{J_1'}\underline{h}^{00}\partial^I L^{J_2'}\partial_t\partial_t h_{\alpha\beta}$:

- When $1\leq|J_1'|\leq N-2$

$$\begin{aligned}&\left\|sL^{J_1'}\underline{h}^{00}\partial^I L^{J_2'}\partial_t\partial_t h_{\alpha\beta}\right\|_{L^2(\mathcal{H}^*_s)}\\&\leq CC_1\varepsilon\left\|s\left(t^{-1}+(s/t)^2t^{-1/2}s^\delta\right)\partial^I L^{J_2'}\partial_t\partial_t h_{\alpha\beta}\right\|_{L^2(\mathcal{H}^*_s)}\\&\leq CC_1\varepsilon\left\|(s/t)\partial^I L^{J_2'}\partial_t\partial_t h_{\alpha\beta}\right\|_{L^2(\mathcal{H}^*_s)}\\&\quad+CC_1\varepsilon s^{1/2+\delta}\left\|(s/t)^{5/2}\partial^I L^{J_2'}\partial_t\partial_t h_{\alpha\beta}\right\|_{L^2(\mathcal{H}^*_s)}\\&\leq CC_1\varepsilon\left\|(s/t)\partial^I L^{J_2'}\partial_t\partial_t h_{\alpha\beta}\right\|_{L^2(\mathcal{H}^*_s)}\\&\quad+CC_1\varepsilon s^{1/2+\delta}\sum_{|J'|<|J|}\left\|(s/t)^{5/2}\partial^I L^{J'}\partial_t\partial_t h_{\alpha\beta}\right\|_{L^2(\mathcal{H}^*_s)}\\&\leq C(C_1\varepsilon)^2s^{-1/2+3\delta}+CC_1\varepsilon\left\|(s/t)\partial^I L^{J_2'}\partial_t\partial_t h_{\alpha\beta}\right\|_{L^2(\mathcal{H}^*_s)}.\end{aligned}$$

- When $|J_1'|\geq N-1$, then $|J_2'|+|I|\leq 1\leq N-4$, we apply (7.1) to $\partial^{J_1'}\underline{h}^{00}$:

$$\begin{aligned}&\left\|sL^{J_1'}\underline{h}^{00}\partial^I L^{J_2'}\partial_t\partial_t h_{\alpha\beta}\right\|_{L^2(\mathcal{H}^*_s)}\\&\leq CC_1\varepsilon\left\|st^{-1}\partial^I L^{J_2'}\partial_t\partial_t h_{\alpha\beta}\right\|_{L^2(\mathcal{H}^*_s)}+\left\|sL^{J_1'}\underline{h}_1^{00}\partial^I L^{J_2'}\partial_t\partial_t h_{\alpha\beta}\right\|_{L^2(\mathcal{H}^*_s)}\\&\leq C(C_1\varepsilon)\left\|(s/t)\partial^I L^{J_2'}\partial_t\partial_t h_{\alpha\beta}\right\|_{L^2(\mathcal{H}^*_s)}\\&\quad+\left\|s^{-1}(s/t)^{-1+\delta}L^{J_1'}\underline{h}_1^{00}\right\|_{L^2(\mathcal{H}^*_s)}\left\|s^2(s/t)^{1-\delta}\partial^I L^{J_2'}\partial_t\partial_t h_{\alpha\beta}\right\|_{L^\infty(\mathcal{H}^*_s)}\\&\leq C(C_1\varepsilon)^2s^{-1/2+3\delta}+CC_1\varepsilon\left\|(s/t)\partial^I L^{J_2'}\partial_t\partial_t h_{\alpha\beta}\right\|_{L^2(\mathcal{H}^*_s)}.\end{aligned}$$

For the term $\underline{h}^{00}\partial_\gamma\partial_{\gamma'}\partial^I L^{J'} h_{\alpha\beta}$, the estimate is similar:

$$
\begin{aligned}
&\left\|s\underline{h}^{00}\partial_\gamma\partial_{\gamma'}\partial^I L^{J'} h_{\alpha\beta}\right\|_{L^2(\mathcal{H}_s^*)} \\
&\leq CC_1\varepsilon\left\|(s/t)\partial_\gamma\partial_{\gamma'}\partial^I L^{J'} h_{\alpha\beta}\right\|_{L^2(\mathcal{H}_s^*)} + \left\|(s/t)^2 t^{-1/2} s^{1+\delta}\partial_\gamma\partial_{\gamma'}\partial^I L^{J'} h_{\alpha\beta}\right\|_{L^2(\mathcal{H}_s^*)} \\
&\leq CC_1\varepsilon\left\|(s/t)\partial_\gamma\partial_{\gamma'}\partial^I L^{J'} h_{\alpha\beta}\right\|_{L^2(\mathcal{H}_s^*)} \\
&+ CC_1\varepsilon s^{1/2+\delta}\sum_{|J'|<|J|}\left\|(s/t)^{5/2}\partial_\gamma\partial_{\gamma'}\partial^I L^{J'} h_{\alpha\beta}\right\|_{L^2(\mathcal{H}_s^*)} \\
&\leq C(C_1\varepsilon)^2 s^{-1/2+3\delta} + CC_1\varepsilon\left\|(s/t)\partial_\gamma\partial_{\gamma'}\partial^I L^{J'} h_{\alpha\beta}\right\|_{L^2(\mathcal{H}_s^*)}.
\end{aligned}
$$

Now, $\left|\partial^I L^J\partial_t\partial_t h_{\alpha\beta}\right| \leq \sum_{\substack{|J'|\leq|J| \\ \gamma,\gamma'}}\left|\partial_\gamma\partial_{\gamma'}\partial^I L^{J'} h_{\alpha\beta}\right|$ in view of the commutator estimates (3.54), and, by the same argument after (7.23),

$$
\left\|(s/t)\partial_\gamma\partial_{\gamma'}\partial^I L^J h_{\alpha\beta}\right\|_{L^2(\mathcal{H}_s^*)} \leq \sum_{|J'|\leq|J|}\left\|(s/t)\partial_t\partial_t\partial^I L^{J'} h_{\alpha\beta}\right\|_{L^2(\mathcal{H}_s^*)} + CC_1\varepsilon s^{-1+\delta}.
$$

So, we conclude that

$$
\begin{aligned}
&\left\|(s/t)\partial^I L^J\partial_t\partial_t h_{\alpha\beta}\right\|_{L^2(\mathcal{H}_s^*)} + \left\|(s/t)\partial_\gamma\partial_{\gamma'}\partial^I L^J h_{\alpha\beta}\right\|_{L^2(\mathcal{H}_s^*)} \\
&\leq C\sum_{|J'|\leq|J|}\left\|s^3 t^{-2}\partial_t\partial_t\partial^I L^{J'} h_{\alpha\beta}\right\|_{L^2(\mathcal{H}_s^*)} + CC_1\varepsilon s^{-1+\delta}. \quad \square
\end{aligned}
$$

Chapter 9

Sup-Norm Estimate Based on Characteristics

9.1 Main statement in this section

Our goal in this section is to control null derivatives, as now stated.

Proposition 9.1. *Assume that* (5.1) *and* (5.2) *hold with $C_1\varepsilon$ sufficiently small, then for $|I|+|J|\leq N-4$,*

$$|(\partial_t-\partial_r)\partial^I L^J\partial_\alpha \underline{h}_{a\beta}|\leq CC_1\varepsilon t^{-1+C\varepsilon}, \tag{9.1}$$

$$|(\partial_t-\partial_r)\partial^I \underline{h}_{a\beta}|\leq CC_1\varepsilon t^{-1}. \tag{9.2}$$

Proof. The proof relies on our earlier estimate along characteristics. We first write the estimate on the components $\underline{h}_{a0}$ in details, and then we sketch the proof on $\underline{h}_{ab}$.

Step I. Estimates for the correction terms. We observe that the equation satisfied by $\underline{h}_{0a}$:

$$\begin{aligned}\widetilde{\Box}_g\underline{h}_{0a} =& \Phi_0^{\alpha'}\Phi_a^{\beta'}Q_{\alpha'\beta'}+\underline{P}_{0a}-16\pi\underline{\partial}_a\phi\partial_t\phi-8\pi\underline{m}_{a0}\phi^2\\ &+\frac{2}{t}\underline{\partial}_a h_{00}-\frac{2x^a}{t^3}h_{00}+Cub(0,0).\end{aligned}$$

Differentiating this equation with respect to $\partial^I L^J$, we have

$$\begin{aligned}&\widetilde{\Box}_g(\partial^I L^J\underline{h}_{0a})\\ &=\partial^I L^J(\Phi_0^{\alpha'}\Phi_a^{\beta'}Q_{\alpha'\beta'})+\partial^I L^J(\underline{P}_{0a})\\ &\quad-16\pi\partial^I L^J(\underline{\partial}_a\phi\partial_t\phi)-8\pi\partial^I L^J(\underline{m}_{a0}\phi^2)\\ &\quad-[\partial^I L^J,h^{\mu\nu}\partial_\mu\partial_\nu]\underline{h}_{a0}+\partial^I L^J\left(\frac{2}{t}\underline{\partial}_a h_{00}-\frac{2x^a}{t^3}h_{00}\right)+\partial^I L^J Cub(0,0).\end{aligned} \tag{9.3}$$

Then we apply Lemma 3.3 to this equation. We need to estimate the L^∞ norm of the terms in the right-hand side and the corrective $M_s[\partial^I L^J\underline{h}_{a0},h]$.

First of all, in view of (6.1), the null terms $\Phi_0^{\alpha'}\Phi_a^{\beta'}Q_{\alpha'\beta'}$ decay like $C(C_1\varepsilon)^2t^{-2}s^{-1+2\delta}$ and in view of (6.2), the quadratic terms QS_ϕ is bounded by $C(C_1\varepsilon)^2t^{-2}s^{-1/2+2\delta}$. We also observe that by the tensorial structure of the Einstein equation, the term $\partial^IL^JP_{a\beta}$ is also a null term, so it is bounded by $C(C_1\varepsilon)^2t^{-2}s^{-1+2\delta}$. We also point out that the high-order terms $\partial^IL^JCub(0,0)$ enjoy also the sufficient decay $C(C_1\varepsilon)^2t^{-2}s^{-1+2\delta}$.

We focus on the linear correction terms $\partial^IL^J\left(\frac{2}{t}\underline{\partial}_a h_{00} - \frac{2x^a}{t^3}h_{00}\right)$. We observe that this term is a linear combination of $t^{-1}\partial^IL^J\underline{\partial}_a h_{00}$ and $t^{-2}\partial^IL^Jh_{00}$ with $|I|+|J|\leq N-4$ with smooth and homogeneous coefficients of degree ≤ 0. Then, these terms can be bounded by $CC_1\varepsilon t^{-5/2}s^{\delta}$.

Then, we analyze the commutator term $[\partial^IL^J, h^{\mu\nu}\partial_\mu\partial_\nu]\underline{h}_{a0}$. We recall that $\underline{h}_{a0}$ is a linear combination of $h_{\alpha\beta}$ with smooth and homogeneous coefficients of degree zero, then the estimate for this term relies on Lemma 4.3. In the list (4.14), we observe that we need only to estimate the terms $\partial^{I_1}L^{J_1}\underline{h}^{00}\partial^{I_2}L^{J_2}\partial_t\partial_t h_{\alpha\beta}$, $L^{J_1'}\underline{h}^{00}\partial^IL^{J_2'}\partial_t\partial_t h_{\alpha\beta}$, $\underline{h}^{00}\partial_\gamma\partial_{\gamma'}\partial^IL^{J'}h_{\alpha\beta}$, since the remaining terms can be bounded by $C(C_1\varepsilon)^2t^{-2}s^{-1+2\delta}$ (see the proof of Lemma 7.6). For the above three terms, we apply (8.15), (8.17) and (7.2):

$$
\begin{aligned}
&\left|L^{J_1'}\underline{h}^{00}\partial^IL^{J_2'}\partial_t\partial_t h_{\alpha\beta}\right| \\
&\leq CC_1\varepsilon\left|\left(t^{-1}+(s/t)^2t^{-1/2}s^{\delta}\right)\partial^IL^{J_2'}\partial_t\partial_t h_{\alpha\beta}\right| \\
&\leq CC_1\varepsilon t^{-1}\left|\partial^IL^{J_2'}\partial_t\partial_t h_{\alpha\beta}\right| + C(C_1\varepsilon)^2t^{-2}s^{-1+3\delta} \\
&\leq CC_1\varepsilon t^{-1}\sum_{\substack{|J_1'|\leq|J'|\\ \gamma,\gamma'}}\left|\partial_\gamma\partial_{\gamma'}\partial^IL^{J_2'}h_{\alpha\beta}\right| + C(C_1\varepsilon)^2t^{-2}s^{-1+3\delta},
\end{aligned}
$$

and

$$
\left|\underline{h}^{00}\partial_\gamma\partial_{\gamma'}\partial^IL^{J'}h_{\alpha\beta}\right| \leq CC_1\varepsilon t^{-1}\left|\partial_\gamma\partial_{\gamma'}\partial^IL^{J'}h_{\alpha\beta}\right| + C(C_1\varepsilon)^2t^{-2}s^{-1+3\delta},
$$

where in the last inequality we applied (8.15). Then thanks to (7.23) and the discussion below these identities in the proof of Lemma 7.6,

$$
\left|\partial_\gamma\partial_{\gamma'}\partial^IL^{J'}h_{\alpha\beta}\right| \leq CC_1\varepsilon t^{-3/2}s^{-1+\delta} + \left|\partial_t\partial_t\partial^IL^{J'}h_{\alpha\beta}\right|,
$$

so that

$$
\left|\underline{h}^{00}\partial_\gamma\partial_{\gamma'}\partial^IL^{J'}h_{\alpha\beta}\right| \leq C(C_1\varepsilon)^2t^{-2}s^{-1+3\delta} + CC_1\varepsilon t^{-1}\left|\partial_t\partial_t\partial^IL^{J'}h_{\alpha\beta}\right|.
$$

Then, by combining this with the commutator estimates, we obtain

$$
\begin{aligned}
&\left|[\partial^IL^J, h^{\mu\nu}\partial_\mu\partial_\nu]\underline{h}_{a0}\right| \\
&\leq Cm_St^{-1}\sum_{|J'|<|J|}\left|\partial^IL^{J'}\partial_\alpha\partial_\beta\underline{h}_{a0}\right| + C(C_1\varepsilon)^2t^{-2}s^{-1+3\delta}.
\end{aligned}
\tag{9.4}
$$

Finally we analyze the correction term $M_s[\partial^I L^J \underline{h}_{a0}, h]$. We recall that

$$M_s[\partial^I L^J \underline{h}_{a0}, h] = r \sum_{a<b} (r\Omega_{ab})^2 u + \underline{h}^{00} W_1[\partial^I L^J \underline{h}_{a0}] + rR[\partial^I L^J \underline{h}_{a0}, h].$$

We see that $r^{-1}\Omega_{ab} = \frac{x^a}{r}\underline{\partial}_b - \frac{x^b}{r}\underline{\partial}_a$ is a linear combination of the "good" terms. So by a similar argument to (4.20), we have

$$\left| \left(r^{-1}\Omega_{ab}\right)^2 \partial^I L^J \underline{h}_{a0}\right| \leq CC_1 \varepsilon t^{-5/2} s^{\delta}.$$

The term W_1 is a linear combination of first- and second-order derivatives with coefficients bounded in $\mathcal{K}\backslash\mathcal{K}^{\text{int}}$. We apply (7.2) to $\underline{h}^{00}$, and we get

$$\left|\underline{h}^{00} W_1[\partial^I L^J \underline{h}_{a0}]\right| \leq C(C_1\varepsilon)^2 t^{-2} s^{2\delta}.$$

The term $R[\partial^I L^J \underline{h}_{a0}, h]$ is bounded similarly, and is a linear combination of the quadratic terms of the following form with homogeneous coefficients: $\underline{h}^{\alpha\beta}\underline{\partial}_a\underline{\partial}_\beta\partial^I L^J \underline{h}_{a0}$, $t^{-1}\underline{h}^{\alpha\beta}\underline{\partial}_\beta\partial^I L^J \underline{h}_{a0}$. For the first term, we apply (4.20) and (5.22): the linear part of $\underline{h}^{\alpha\beta}$ is a linear combination of $h_{\alpha\beta}$ with smooth and homogeneous coefficients of degree zero. The second term is bounded by the additional decreasing factor t^{-1} and therefore

$$\left|R[\partial^I L^J \underline{h}_{a0}, h]\right| \leq C(C_1\varepsilon)^2 t^{-3} s^{2\delta}.$$

Then we conclude that

$$|M_s[\partial^I L^J \underline{h}_{a0}, h](t,x)| \leq CC_1\varepsilon t^{-3/2} s^{2\delta}, \quad 3/5 \leq r/t \leq 1, \ |I| + |J| \leq N-4.$$

Step II. Case of $|J| = 0$. Now we substitute the above estimate into the inequality (3.18) and observe that when $|J| = 0$, the first term in the right-hand side of (9.4) disappears. Then, we have

$$\begin{aligned}
|(\partial_t - \partial_r)\partial^I \underline{h}_{a0}| &\leq Ct^{-1} \sup_{\partial_B\mathcal{K}^{\text{int}}_{[2,s^*]}\cup\partial\mathcal{K}} \{|(\partial_t - \partial_r)(r\partial^I \underline{h}_{a0})|\} + Ct^{-1}|\partial^I \underline{h}_{a0}(t,x)| \\
&\quad + C(C_1\varepsilon)^2 t^{-1}\int_{a_0}^t \tau^{-5/4+3\delta}d\tau + CC_1\varepsilon t^{-1}\int_{a_0}^t \tau^{-3/2+3\delta}d\tau \\
&\leq CC_1\varepsilon t^{-1} + Ct^{-1} \sup_{\partial_B\mathcal{K}^{\text{int}}_{[2,s_0]}\cup\partial\mathcal{K}} \{|(\partial_t - \partial_r)(r\partial^I \underline{h}_{a0})|\}.
\end{aligned}$$

Observe that on the boundary $\partial_B\mathcal{K}^{\text{int}}_{[2,s_0]}$, $r = 3t/5$. We have

$$\begin{aligned}
&|(\partial_t - \partial_r)(r\partial^I \underline{h}_{a0})| \leq r|(\partial_r - \partial_t)\partial^I \underline{h}_{a0}| + |\partial^I \underline{h}_{a0}| \\
&\leq CC_1\varepsilon r t^{-1/2} s^{-1+\delta} + Cm_S\varepsilon t^{-1} + CC_1\varepsilon(s/t)t^{-1/2}s^{\delta} \\
&\leq CC_1\varepsilon r t^{-3/2+\delta/2} + CC_1\varepsilon t^{-1} + CC_1\varepsilon(s/t)t^{-1/2}s^{\delta} \leq CC_1\varepsilon.
\end{aligned}$$

We also observe that on $\partial\mathcal{K}$, $\underline{h}_{a0} = \underline{h}_{sa0}$,

$$|(\partial_t - \partial_r)(r\partial^I \underline{h}_{a0})| \leq r|(\partial_r - \partial_t)\underline{h}_{a0}| + |\underline{h}_{a0}| \leq Cm_S\varepsilon r t^{-1} + Cm_S\varepsilon t^{-1}$$
$$\leq CC_1\varepsilon.$$

This leads us to (9.2) for $\underline{h}_{0a}$.

Step III. Induction on $|J|$. The proof of (9.1) is done by induction on $|J|$. The initial case $|J| = 0$ is already guaranteed in view of (9.2). We assume that (9.1) holds for all $0 \leq |J'| \leq n < N - 4$ and we will prove it with $|J| = n$. First, based on (9.1), the following result is immediate:

$$|\partial_\alpha \partial^I L^J \underline{h}_{a0}| + |\partial^I L^I \partial_\alpha \underline{h}_{0a}| \leq CC_1\varepsilon t^{-1+C\varepsilon}, \quad |I| + |J| \leq N - 4, \tag{9.5}$$

$$|\partial_\alpha \partial^I \underline{h}_{a0}| \leq CC_1\varepsilon t^{-1}, \quad |I| \leq N - 4. \tag{9.6}$$

These are based on the identity $\partial_t = \frac{t-r}{t}\partial_t + \frac{x^a}{t+r}\underline{\partial}_a + \frac{r}{t+r}(\partial_t - \partial_r)$, where ∂_t can be expressed by the "good" derivatives and $\partial_t - \partial_r$. Furthermore, we have $\partial_a = \underline{\partial}_a - \frac{x^a}{t}\partial_t$ and, then, based on the basic L^∞ estimate of the "good" derivatives and (9.1) and (9.2), the derivation of (9.5) and (9.6) is immediate.

Then we substitute the above estimates on the source terms and corrective term into (3.18). Observe that by the inductive assumption, (9.4) becomes

$$|[\partial^I L^J, \underline{h}^{00}\partial_t\partial_t]\underline{h}_{a0}| \leq C(C_1\varepsilon)^2 t^{-2} s^{-1+3\delta} + C(C_1\varepsilon)^2 t^{-2+C\varepsilon},$$

where we have noticed that $\sum_{|J'|<|J|} |\partial^I L^J \partial_\alpha \partial_\beta \underline{h}_{a0}| \leq CC_1\varepsilon s^{-1+C\varepsilon}$ (by the commutator estimates and (9.5)). This leads us to (in view of (3.18))

$$\begin{aligned}
&|(\partial_t - \partial_r)\partial^I L^J \underline{h}_{a0}| \\
&\leq Ct^{-1} \sup_{\partial_B\mathcal{K}^{\text{int}}_{[2,s^*]} \cup \partial\mathcal{K}} \{|(\partial_t - \partial_r)(r\partial^I L^J \underline{h}_{a0})|\} + Ct^{-1}|\partial^I L^J \underline{h}_{a0}(t,x)| \\
&\quad + C(C_1\varepsilon)^2 t^{-1} \int_{a_0}^t \tau^{-1+C\varepsilon} d\tau + CC_1\varepsilon t^{-1} \int_{a_0}^t \tau^{-3/2+2\delta} d\tau \\
&\leq CC_1\varepsilon t^{-1+C\varepsilon} + Ct^{-1} \sup_{\partial_B\mathcal{K}^{\text{int}}_{[2,s_0]} \cup \partial\mathcal{K}} \{|(\partial_t - \partial_r)(r\underline{h}_{a0})|\}.
\end{aligned}$$

Then, similarly as in the argument above, (9.1) is proved for $\underline{h}_{0a}$.

The estimate for $\underline{h}_{ab}$ is similar, where we also observe that the quasi-null terms $\underline{P}_{ab}$ are eventually null terms, and the correction terms behave the same decay as in the case of $\underline{h}_{a0}$. □

9.2 Application to quasi-null terms

Our main application of the refined sup-norm estimate concerns the terms $P_{\alpha\beta}$.

Lemma 9.1. *Let (I, J) be a multi-index and $|I| + |J| \leq N$. Then, one has*

$$\begin{aligned}
&\|\partial^I L^J P_{\alpha\beta}\|_{L^2(\mathcal{H}_s^*)} \leq CC_1\varepsilon s^{-1} \sum_{\alpha',\beta'} E_M^*(s, \partial^I L^J h_{\alpha'\beta'})^{1/2} \\
&+ CC_1\varepsilon s^{-1} \sum_{\substack{|I'|<|I| \\ \alpha',\beta'}} E_M^*(s, \partial^{I'} L^J h_{\alpha'\beta'})^{1/2} \\
&+ CC_1\varepsilon s^{-1+CC_1\varepsilon} \sum_{\substack{|I'|\leq|I|,|J'|<|J| \\ \alpha',\beta'}} E_M^*(s, \partial^{I'} L^{J'} h_{\alpha'\beta'})^{1/2} + C(C_1\varepsilon)^2 s^{-3/2+2\delta}.
\end{aligned} \tag{9.7}$$

Proof. We apply Lemma 4.9 combined with the estimates (9.5) and (9.6). We first observe that due to its tensorial structure, the estimate for $P_{\alpha\beta}$ can be relined on the estimates on $\underline{P}_{\alpha\beta}$. Furthermore, the components $\underline{P}_{a\beta}$ or $\underline{P}_{\alpha b}$ are essentially null terms (see (4.38)), so that $\|\partial^I L^J \underline{P}_{a\beta}\|_{L^2(\mathcal{H}_s^*)} \leq C(C_1\varepsilon)^2 s^{-3/2+2\delta}$. We focus on $\underline{P}_{00}$. We see that in the list (4.37), the non-trivial term are linear combinations of $\partial_t \underline{h}_{a\alpha} \partial_t \underline{h}_{b\beta}$ with smooth and homogeneous coefficients of degree zero. Then we only need to estimate $\|\partial^I L^J (\partial_t \underline{h}_{a\alpha} \partial_t \underline{h}_{b\beta})\|_{L^2(\mathcal{H}_s^*)}$ for $|I| + |J| \leq N$. We have

$$\|\partial^I L^J (\partial_t \underline{h}_{a\alpha} \partial_t \underline{h}_{b\beta})\|_{L^2(\mathcal{H}_s^*)} \leq \sum_{\substack{I_1+I_2=I \\ J_1+J_2=J}} \|\partial^{I_1} L^{J_1} \partial_t \underline{h}_{a\alpha}\, \partial^{I_2} L^{J_2} \partial_t \underline{h}_{b\beta}\|_{L^2(\mathcal{H}_s^*)}.$$

Recall that $N \geq 7$ then either $|I_1| + |J_1| \leq N - 4$ or $|I_2| + |J_2| \leq N - 4$. Without loss of generality, we suppose that $|I_1| + |J_1| \leq N - 4$. Then

- When $J_1 = 0$, we apply (9.6):

$$\begin{aligned}
&\|\partial^{I_1} \partial_t \underline{h}_{a\alpha}\, \partial^{I_2} L^J \partial_t \underline{h}_{b\beta}\|_{L^2(\mathcal{H}_s^*)} \leq CC_1\varepsilon \|t^{-1} \partial^{I_2} L^J \partial_t \underline{h}_{b\beta}\|_{L^2(\mathcal{H}_s^*)} \\
&\leq CC_1\varepsilon s^{-1} \|(s/t)\partial^{I_2} L^J \partial_t \underline{h}_{b\beta}\|_{L^2(\mathcal{H}_s^*)} \\
&\leq CC_1\varepsilon s^{-1} \sum_{\substack{|I'|\leq|I|,|J'|\leq|J| \\ \gamma,\gamma'}} E_M^*(s, \partial^{I'} L^{J'} h_{\gamma\gamma'})^{1/2}.
\end{aligned}$$

- When $|J_1| \geq 1, 1 \leq |I_1| + |J_1| \leq N - 4$, we apply (9.5):

$$
\begin{aligned}
&\|\partial^{I_1} L^{J_1} \partial_t \underline{h}_{a\alpha}\, \partial^{I_2} L^{J_2} \partial_t \underline{h}_{b\beta}\|_{L^2(\mathcal{H}_s^*)} \\
&\leq CC_1 \varepsilon s^{-1+CC_1\varepsilon} \|(s/t) \partial^{I_2} L^{J_2} \partial_t \underline{h}_{b\beta}\|_{L^2(\mathcal{H}_s^*)} \\
&\leq CC_1 \varepsilon s^{-1+CC_1\varepsilon} \sum_{\substack{|I'| \leq |I_2|, |J'| \leq |J_2| \\ \alpha, \beta}} E_M^*(s, \partial^{I'} L^{J'} h_{\gamma,\gamma'})^{1/2} \\
&\leq CC_1 \varepsilon s^{-1+CC_1\varepsilon} \sum_{\substack{|I'| \leq |I_2|, |J'| < |J| \\ \alpha, \beta}} E_M^*(s, \partial^{I'} L^{J'} h_{\gamma,\gamma'})^{1/2}. \quad \square
\end{aligned}
$$

Chapter 10

Low-Order Refined Energy Estimate for the Spacetime Metric

10.1 Preliminary

In this section, we improve the energy bounds on $E_M^*(s, \partial^I L^J h_{\alpha\beta})$ for $|I| + |J| \leq N - 4$. We apply Proposition 3.1. In this case the L^2 norm of $\partial^I L^J \left(\partial_\alpha \phi \partial_\beta \phi + \phi^2\right)$ is integrable with respect to s. We need to focus on the estimate of $F_{\alpha\beta}$ and the commutators $[\partial^I L^J, h^{\mu\nu}\partial_\mu\partial_\nu]h_{\alpha\beta}$.

Lemma 10.1. *Under the bootstrap assumption* (5.1) *and* (5.2) *with* $C_1\varepsilon$ *sufficiently small, one has for* $|I| + |J| \leq N$:

$$
\begin{aligned}
&\left\|\partial^I L^J F_{\alpha\beta}\right\|_{L^2(\mathcal{H}_s^*)} \\
&\leq C(C_1\varepsilon)^2 s^{-3/2+2\delta} + CC_1\varepsilon s^{-1} \sum_{\alpha',\beta'} E_M^*(s, \partial^I L^J h_{\alpha'\beta'})^{1/2} \\
&\quad + CC_1\varepsilon s^{-1} \sum_{\substack{|I'|<|I| \\ \alpha',\beta'}} E_M^*(s, \partial^{I'} L^J h_{\alpha'\beta'})^{1/2} \\
&\quad + CC_1\varepsilon s^{-1+CC_1\varepsilon} \sum_{\substack{|I'|\leq|I|,|J'|<|J| \\ \alpha',\beta'}} E_M^*(s, \partial^{I'} L^{J'} h_{\alpha'\beta'})^{1/2}.
\end{aligned}
\tag{10.1}
$$

Proof. We use here (9.7). We observe that $F_{\alpha\beta} = Q_{\alpha\beta} + P_{\alpha\beta}$, where $Q_{\alpha\beta}$ are null terms combined with higher-order (cubic) terms. Then trivial substitution of the basic L^2 and sup-norm estimates (see the proof of (6.7)) shows that $\left\|\partial^I L^J Q_{\alpha\beta}\right\|_{L^2(\mathcal{H}_s^*)} \leq C(C_1\varepsilon)^2 s^{-3/2+2\delta}$. The estimate for $P_{\alpha\beta}$ is provided by (9.7). □

Lemma 10.2. *Under the bootstrap assumption* (5.1) *and* (5.2), *the following* *estimates holds for* $|I| + |J| \leq N - 4$*:*

$$
\begin{aligned}
&\big\|[\partial^I L^J, h^{\mu\nu}\partial_\mu\partial_\nu]h_{\alpha\beta}\big\|_{L^2(\mathcal{H}^*_s)} \\
&\leq C(C_1\varepsilon)^2 s^{-3/2+2\delta} + CC_1\varepsilon s^{-1} \sum_{a,|J'|<|J|} E^*_M(s, \partial^I L_a L^{J'} h_{\alpha\beta})^{1/2} \\
&\quad + CC_1\varepsilon s^{-1+CC_1\varepsilon} \sum_{\substack{|I'|\leq|I| \\ |J'|<|J|}} \sum_{\alpha',\beta'} E^*_M(s, \partial^{I'} L^{J'} h_{\alpha'\beta'})^{1/2}.
\end{aligned}
\tag{10.2}
$$

Proof. This is based on (8.20). We need to estimate the term $\Big\|(s/t)^2\partial_t\partial_t\partial^I L^{J'} h_{\alpha\beta}\Big\|_{L^2(\mathcal{H}^*_s)}$ with $|J'| < |J|$. We are going to use (8.5). We see that in view of (8.10):

$$
\begin{aligned}
&\Big\|Sc_1[\partial^I L^{J'} h_{\alpha\beta}]\Big\|_{L^2(\mathcal{H}^*_s)} \\
&\leq Cs^{-1}\sum_a E^*_M(s, \partial^I L_a L^{J'} h_{\alpha\beta})^{1/2} + Cs^{-1} \sum_{|I'|\leq|I|} E^*_M(s, \partial^{I'} L^{J'} h_{\alpha\beta})^{1/2}.
\end{aligned}
$$

The term Sc_2 is bounded in view of (8.12):

$$
\Big\|Sc_2[\partial^I L^{J'} h_{\alpha\beta}]\Big\|_{L^2(\mathcal{H}^*_s)} \leq C(C_1\varepsilon)^2 s^{-3/2+2\delta}.
$$

The term $F_{\alpha\beta}$ is bounded by Lemma 10.1.

For the term QS_ϕ, we will only analyze in detail the term $\partial_\alpha\phi\partial_\beta\phi$ and omit the proof on ϕ^2. We see first that $\partial^I L^{J'}(\partial_\alpha\phi\partial_\beta\phi) = \sum_{\substack{I_1+I_2=I \\ J_1+J_2=J'}} \partial^{I_1}L^{J_1}\partial_\alpha\phi\,\partial^{I_2}L^{J_2}\partial_\beta\phi$. We then observe that, for $N \geq 7$ and $|I| + |J'| \leq N - 5$, either $|I_1| + |J_1| \leq N - 6$ or $|I_2| + |J_2| \leq N - 6$. Suppose without loss of generality that $|I_1| + |J_1| \leq N - 6$. Then we have $\Big\|\partial^I L^{J'}(\partial_\alpha\phi\partial_\beta\phi)\Big\|_{L^2(\mathcal{H}^*_s)} \leq \big\|\partial^{I_1}L^{J_1}\partial_\alpha\phi\,\partial^{I_2}L^{J_2}\partial_\beta\phi\big\|_{L^2(\mathcal{H}^*_s)}$.

- when $I_1 = J_1 = 0$, we see that $0 \leq N - 7$, then we have

$$
\begin{aligned}
\Big\|\partial^I L^{J'}(\partial_\alpha\phi\partial_\beta\phi)\Big\|_{L^2(\mathcal{H}^*_s)} &\leq \big\|(t/s)\partial_\alpha\phi\,(s/t)\partial^{I_2}L^{J_2}\partial_\beta\phi\big\|_{L^2(\mathcal{H}^*_s)} \\
&\leq CC_1\varepsilon\Big\|(t/s)t^{-3/2}s^\delta\,(s/t)\partial^{I_2}L^{J_2}\partial_\beta\phi\Big\|_{L^2(\mathcal{H}^*_s)} \\
&\leq CC_1\varepsilon s^{-3/2+\delta}\big\|(s/t)\partial^{I_2}L^{J_2}\partial_\beta\phi\big\|_{L^2(\mathcal{H}^*_s)} \\
&\leq C(C_1\varepsilon)^2 s^{-3/2+2\delta}.
\end{aligned}
$$

- when $1 \leq |I_1| + |I_2| \leq N - 6$, we see that $|I_2| + |J_2| \leq N - 5$. So we have

$$
\begin{aligned}
\Big\|\partial^I L^{J'}(\partial_\alpha\phi\partial_\beta\phi)\Big\|_{L^2(\mathcal{H}^*_s)} &\leq \big\|\partial^{I_1}L^{J_1}\partial_\alpha\phi\big\|_{L^\infty(\mathcal{H}^*_s)}\big\|\partial^{I_2}L^{J_2}\partial_\beta\phi\big\|_{L^2(\mathcal{H}^*_s)} \\
&\leq CC_1\varepsilon s^{-3/2}\,CC_1\varepsilon s^\delta \leq C(C_1\varepsilon)^2 s^{-3/2+2\delta}.
\end{aligned}
$$

We conclude that

$$\left\|QS_\phi(p,k)\right\|_{L^2(\mathcal{H}_s^*)} \leq C(C_1\varepsilon)^2 s^{-3/2+2\delta}, \qquad p \leq N-4. \tag{10.3}$$

The term $[\partial^I L^{J'}, h^{\mu\nu}\partial_\mu\partial_\nu]h_{\alpha\beta}$ is conserved. Then we see the following estimate are established:

$$\begin{aligned}
&\left\|[\partial^I L^J, h^{\mu\nu}]h_{\alpha\beta}\right\|_{L^2(\mathcal{H}_s^*)} \\
&\leq CC_1\varepsilon s^{-1} \sum_{\substack{\alpha',\beta',a\\|J'|<|J|}} E_M^*(s, \partial^I L_a L^{J'} h_{\alpha'\beta'})^{1/2} \\
&\quad + CC_1\varepsilon s^{-1+CC_1\varepsilon} \sum_{\substack{\alpha',\beta'\\|I'|\leq|I|\\|J'|<|J|}} E_M^*(s, \partial^{I'} L^{J'} h_{\alpha'\beta'})^{1/2} \\
&\quad + \sum_{\substack{\alpha',\beta'\\|J'|<|J|}} \left\|[\partial^I L^{J'}, h^{\mu\nu}\partial_\mu\partial_\nu]h_{\alpha'\beta'}\right\|_{L^2_f(\mathcal{H}_s)} + C(C_1\varepsilon)^2 s^{-3/2+2\delta}.
\end{aligned} \tag{10.4}$$

We proceed by induction on $|J|$. In (10.4), if we take $|J| = 0$, then only the last term in the right-hand side exists, this concludes (10.2). Assume that (10.2) holds for $|J| \leq n-1 \leq N-5$, we will prove that it still holds for $|J| = n \leq N-4$. We substitute (10.2) into the last term in the right-hand side of (10.4). □

10.2 Main estimate established in this section

Proposition 10.1 (Lower order refined energy estimate for $h_{\alpha\beta}$). *There exists a constant $\varepsilon_1 > 0$ determined by $C_1 > 2C_0$ such that assume that the bootstrap assumption (5.1) holds with (C_1, ε), $0 \leq \varepsilon \leq \varepsilon_1$, then the following refined estimate holds*

$$E_M(s, \partial^I L^J h_{\alpha\beta})^{1/2} \leq \frac{1}{2} C_1 \varepsilon s^{CC_1\varepsilon}, \quad \alpha,\beta \leq 3, \quad |I|+|J| \leq N-4. \tag{10.5}$$

Proof. The proof relies on a direct application of Proposition 3.1. We need to bound the terms presented in the right-hand side of (3.2). The term $F_{\alpha\beta}$ is bounded by Lemma 10.1, the term QS_ϕ is bounded in view of (10.3). The estimate for $[\partial^I L^J, h^{\mu\nu}\partial_\mu\partial_\nu]h_{\alpha\beta}$ is obtained in view of (10.2). By (7.6), the term $M_{\alpha\beta}[\partial^I L^J h]$ is bounded by $C(C_1\varepsilon)^2 s^{-3/2+2\delta}$. Then in view of (3.2):

$$
\begin{aligned}
&\sum_{\alpha,\beta} E_M(s,\partial^I L^J h_{\alpha\beta})^{1/2} \\
&\leq CC_0\,\varepsilon + C(C_1\varepsilon)^2 + CC_1\varepsilon \sum_{\alpha,\beta}\int_2^s \tau^{-1} E^*_M(\tau,\partial^I L^J h_{\alpha\beta})^{1/2} d\tau \\
&+ CC_1\varepsilon \sum_{\substack{|I'|<|I| \\ \alpha,\beta}} \int_2^s \tau^{-1} E^*_M(\tau,\partial^{I'} L^J h_{\alpha\beta})^{1/2} d\tau \\
&+ CC_1\varepsilon \sum_{\substack{|I'|\leq|I|,|J'|<|J| \\ \alpha,\beta}} \int_2^s \tau^{-1+CC_1\varepsilon} E^*_M(\tau,\partial^{I'} L^{J'} h_{\alpha\beta})^{1/2} d\tau \\
&+ CC_1\varepsilon \sum_{\substack{\alpha,\beta,a \\ |J'|<|J|}} \int_2^s \tau^{-1} E^*_M(\tau,\partial^I L_a L^{J'} h_{\alpha\beta})^{1/2} d\tau.
\end{aligned}
\tag{10.6}
$$

The rest of the proof is based on (10.6). When $|J| = 0$, the last two terms in the right-hand side of (10.6) disappears. Then, we have

$$
\begin{aligned}
&\sum_{\substack{\alpha,\beta \\ |I|\leq N-4}} E_M(s,\partial^I h_{\alpha\beta})^{1/2} \\
&\leq C\left(C_0\,\varepsilon + (C_1\varepsilon)^2\right) + CC_1\varepsilon \sum_{\substack{\alpha,\beta \\ |I|\leq N-4}} \int_2^s \tau^{-1} E_M(\tau,\partial^I h_{\alpha\beta})^{1/2} d\tau.
\end{aligned}
$$

Then by Gronwall's inequality, we have

$$
\sum_{\substack{\alpha,\beta \\ |I|\leq N-4}} E_M(s,\partial^I h_{\alpha\beta})^{1/2} \leq C\big(C_0\,\varepsilon + (C_1\varepsilon)^2\big) s^{CC_1\varepsilon}. \tag{10.7}
$$

Here we can already ensure that $\sum_{\alpha,\beta} E_M(s,\partial^I h_{\alpha\beta})^{1/2} \leq \frac{1}{2} C_1\varepsilon s^{CC_1\varepsilon}$ by choosing $\varepsilon_{10} = \frac{C_1 - 2CC_0}{2C_1^2}$ with C_1 sufficiently large.

We proceed by induction on $|J|$ and suppose that

$$
\sum_{\substack{\alpha,\beta \\ |I|\leq N-4}} E_M(s,\partial^I h_{\alpha\beta})^{1/2} \leq C\big(C_0\,\varepsilon + (C_1\varepsilon)^2\big) s^{CC_1\varepsilon} \tag{10.8}
$$

holds for $|J| < n \leq N-4$, we will prove that it still holds for $|J| = n$. Substitute (10.8) into the last two terms of the right-hand side of (10.6), we see that

$$\begin{aligned}
\sum_{\alpha,\beta} E_M(s, \partial^I L^J h_{\alpha\beta})^{1/2} &\le CC_0\,\varepsilon + C(C_1\varepsilon)^2 \\
&\quad + CC_1\varepsilon \sum_{\alpha,\beta} \int_2^s \tau^{-1} E_M(\tau, \partial^I L^J h_{\alpha\beta})^{1/2} d\tau \\
&\quad + CC_1\varepsilon \sum_{\substack{\alpha,\beta \\ |I'|<|I|}} \int_2^s \tau^{-1} E_M(\tau, \partial^{I'} L^J h_{\alpha\beta})^{1/2} d\tau \\
&\quad + CC_1\varepsilon \left(C_0\,\varepsilon + (C_1\varepsilon)^2\right) \int_2^s \tau^{-1+CC_1\varepsilon} d\tau \\
&\quad + CC_1\varepsilon \sum_{\substack{a,\alpha,\beta \\ |J'|=|J|-1}} \int_2^s \tau^{-1} E_M^*(\tau, \partial^I L_a L^{J'} h_{\alpha\beta})^{1/2} d\tau,
\end{aligned}$$

thus

$$\begin{aligned}
&\sum_{\alpha,\beta} E_M(s, \partial^I L^J h_{\alpha\beta})^{1/2} \\
&\le C\left(C_0 + (C_1\varepsilon)^2\right) s^{CC_1\varepsilon} + CC_1\varepsilon \sum_{\alpha,\beta} \int_2^s \tau^{-1} E_M(\tau, \partial^I L^J h_{\alpha\beta})^{1/2} d\tau \\
&\quad + CC_1\varepsilon \sum_{\substack{\alpha,\beta \\ |I'|<|I|}} \int_2^s \tau^{-1} E_M(\tau, \partial^{I'} L^J h_{\alpha\beta})^{1/2} d\tau \\
&\quad + CC_1\varepsilon \sum_{\substack{\alpha,\beta \\ |J'|=|J|}} \int_2^s \tau^{-1} E_M^*(\tau, \partial^I L^{J'} h_{\alpha\beta})^{1/2} d\tau.
\end{aligned}$$

This leads us to

$$\begin{aligned}
&\sum_{\substack{\alpha,\beta,|J|=n \\ |I|\le N-4-n}} E_M(s, \partial^I L^J h_{\alpha\beta})^{1/2} \\
&\le C\left(C_0\,\varepsilon + (C_1\varepsilon)^2\right) s^{CC_1\varepsilon} + CC_1\varepsilon \sum_{\substack{\alpha,\beta,|J|=n \\ |I|\le N-4-n}} \int_2^s \tau^{-1} E_M(\tau, \partial^I L^J h_{\alpha\beta})^{1/2} d\tau.
\end{aligned}$$

Then by Gronwall's inequality, we have (by taking some constant C larger than the one provided the above estimate)

$$\sum_{\substack{\alpha,\beta \\ |I|\le N-4-|J|}} E_M(s, \partial^I L^J h_{\alpha\beta})^{1/2} \le C\left(C_0\,\varepsilon + (C_1\varepsilon)^2\right) s^{CC_1\varepsilon}.$$

By choosing $\varepsilon_{1n} = \frac{C_1 - 2CC_0}{2C_1^2}$, we see that

$$\sum_{\substack{\alpha,\beta \\ |I|\le N-4-|J|}} E_M(s, \partial^I L^J h_{\alpha\beta})^{1/2} \le \frac{1}{2} C_1 \varepsilon s^{CC_1\varepsilon}.$$

Then, we choose $\varepsilon_1 = \min_{0\leq n\leq N-4}\{\varepsilon_{1n}\}$ and conclude that for $\varepsilon \leq \varepsilon_1$, (10.5) is thus proven. □

10.3 Application of the refined energy estimate

The improved low-order energy estimates on $h_{\alpha\beta}$ will lead us to a series of estimates. Based on (10.1), the sup-norm estimates are direct by the global Sobolev inequality (for $|I|+|J|\leq N-6$):

$$|\partial^I L^J \partial_\gamma h_{\alpha\beta}| + |\partial_\gamma \partial^I L^J h_{\alpha\beta}| \leq CC_1\varepsilon t^{-1/2} s^{-1+CC_1\varepsilon}, \tag{10.9}$$

$$|\partial^I L^J \underline{\partial}_a h_{\alpha\beta}| + |\underline{\partial}_a \partial^I L^J h_{\alpha\beta}| \leq CC_1\varepsilon t^{-3/2} s^{CC_1\varepsilon}. \tag{10.10}$$

Based on this improved sup-norm estimate, the following estimates are direct by integration along the radial rays $\{(t,\lambda x)|1\leq\lambda\leq t/|x|\}$:

$$|\partial^I L^J h_{\alpha\beta}| \leq CC_1\varepsilon \left(t^{-1} + (s/t)t^{1/2} s^{CC_1\varepsilon}\right). \tag{10.11}$$

We take the above bounds and substitute them into the proof of Lemma 4.7, following exactly the same procedure, we obtain for $|I|+|J|\leq N-6$:

$$\left|\partial^I L^J \partial_\alpha \underline{h}^{00}\right| + \left|\partial^I L^J \partial_\alpha \underline{h}^{00}\right| \leq CC_1\varepsilon t^{-3/2} s^{CC_1\varepsilon} \tag{10.12}$$

and also by integration along the rays $\{(t,\lambda x)|1\leq\lambda\leq t/|x|\}$ (and taking into account the exterior Schwarzschild metric):

$$\left|\partial^I L^J \underline{h}^{00}\right| \leq CC_1\varepsilon \left(t^{-1} + (s/t)^2 t^{1/2} s^{CC_1\varepsilon}\right). \tag{10.13}$$

Two more delicate applications of this improved energy estimate for $h_{\alpha\beta}$ are now obtained. We begin with $F_{\alpha\beta}$, in view of (10.9).

Lemma 10.3. *For $|I|+|J|\leq N-6$, one has*

$$|\partial^I L^J F_{\alpha\beta}| \leq C(C_1\varepsilon)^2 t^{-2+CC_1\varepsilon}(t-r)^{-1+CC_1\varepsilon}. \tag{10.14}$$

Proof. Observe that $F_{\alpha\beta}$ is a linear combination of GQS_h and $P_{\alpha\beta}$ and in $P_{\alpha\beta}$ the only term to be concerned about (by Lemma 4.9) is $\underline{m}^{0a}\underline{m}^{0b}\partial_t\underline{h}_{0a}\partial_t\underline{h}_{0b}$, the remaining terms are GQS_h, Cub or Com which have better decay. We observe that in view of (10.9),

$$\left|\partial^I L^J \left(\partial_t \underline{h}_{a\alpha}\partial_t \underline{h}_{b\beta}\right)\right| \leq C(C_1\varepsilon)^2 t^{-1} s^{-2+CC_1\varepsilon}. \quad \square$$

Then, a second refined estimate can be established.

Lemma 10.4. *For $|I|+|J|\leq N-7$, one has*

$$\left|\partial_t\partial_t\partial^I L^J h_{\alpha\beta}\right| \leq CC_1\varepsilon t^{1/2} s^{-3+CC_1\varepsilon}. \tag{10.15}$$

Proof. The proof is essentially a refinement of the proof of (8.7). We see that when the energy is improved, in view of (10.9), $|Sc_1[\partial^I L^J h_{\alpha\beta}]|$ is bounded by $CC_1\varepsilon t^{-3/2}s^{-1+CC_1\varepsilon}$ (in view of (8.6)). The term $F_{\alpha\beta}$ is bounded by the above estimate (10.14). The terms Sc_2, QS_ϕ and the commutator are bounded as in the proof of (8.7). Then we get the following estimate parallel to (8.8):

$$
\begin{aligned}
|(s/t)^2\partial_t\partial_t\partial^I L^J h_{\alpha\beta}| \leq{}& CC_1\varepsilon t^{-3/2}s^{-1+CC_1\varepsilon} + C(C_1\varepsilon)^2 t^{-1}s^{-2+CC_1\varepsilon} \\
&+ CC_1\varepsilon\left(t^{-1} + (s/t)^2 t^{-1/2}s^{\delta}\right)\sum_{|J'|<|J|}\left|\partial_t\partial_t\partial^I L^{J'} h_{\alpha\beta}\right|.
\end{aligned}
$$

By induction, the desired result is thus established. □

Proof. The proof is essentially [illegible] argument of the proof [illegible]. [illegible] when the energy [illegible] in view of [illegible] bounded [illegible] be made by the above [illegible]. The [illegible] constants for the bounded [illegible]. Then [illegible] estimate parallel to [illegible]

[illegible]

[illegible] and the [illegible] condition [illegible]

Chapter 11

Low-Order Refined Sup-Norm Estimate for the Metric and Scalar Field

11.1 Main estimates established in this section

Our aim in this section is to establish the estimates: $|I| + |J| \leq N - 7$:

$$|L^J h_{\alpha\beta}| \leq CC_1\varepsilon t^{-1} s^{C(C_1\varepsilon)^{1/2}}, \tag{11.1}$$

$$(s/t)^{3\delta-2}|\partial^I L^J \phi| + (s/t)^{3\delta-3}|\partial^I L^J \underline{\partial}_\perp \phi| \leq CC_1\varepsilon s^{-3/2+C(C_1\varepsilon)^{1/2}}, \tag{11.2}$$

$$(s/t)^{3\delta-2}|\partial^I \phi| + (s/t)^{3\delta-3}|\underline{\partial}_\perp \partial^I \phi| \leq CC_1\varepsilon s^{-3/2}. \tag{11.3}$$

Let us first point out some direct consequences of these three estimates, by noting the relations $\partial_t = (s/t)^{-2}\left(\underline{\partial}_\perp - \frac{x^a}{t}\underline{\partial}_a\right)$ and $\partial_a = \underline{\partial}_a - \frac{x^a}{t}\partial_t$ and the sharp decay rate on $\underline{\partial}_a$ (for $|I| + |J| \leq N - 7$)

$$|\underline{\partial}_a \partial^I L^J \phi(t,x)| \leq CC_1\varepsilon t^{-5/2} s^{1/2+\delta}.$$

So, (11.1), (11.2) and (11.3) lead to

$$\left|\partial_\alpha \partial^I L^J \phi(t,x)\right| \leq CC_1\varepsilon (s/t)^{1-3\delta} s^{-3/2+C(C_1\varepsilon)^{1/2}}, \quad |I| + |J| \leq N - 7, \tag{11.4}$$

$$\left|\partial_\alpha \partial^I L^J \phi(t,x)\right| \leq CC_1\varepsilon (s/t)^{2-3\delta} s^{-3/2+C(C_1\varepsilon)^{1/2}}, \quad |I| + |J| \leq N - 8. \tag{11.5}$$

We also have

$$\left|\partial_\alpha \partial^I \phi(t,x)\right| \leq CC_1\varepsilon (s/t)^{1-3\delta} s^{-3/2}, \quad |I| \leq N - 7, \tag{11.6}$$

$$\left|\partial_\alpha \partial^I \phi(t,x)\right| \leq CC_1\varepsilon (s/t)^{2-3\delta} s^{-3/2}, \quad |I| \leq N - 8. \tag{11.7}$$

In particular, we see that

$$|\partial_\alpha \phi(t,x)| \leq CC_1\varepsilon (s/t)^{2-3\delta} s^{-3/2}. \tag{11.8}$$

We observe that by the commutator estimates:

$$\begin{aligned}
&\left|\partial^I L^J \partial_\alpha \phi\right| \leq CC_1\varepsilon (s/t)^{1-3\delta} s^{-3/2+C(C_1\varepsilon)^{1/2}}, && |I| + |J| \leq N - 7,\\
&\left|\partial^I L^J \partial_\alpha \phi\right| \leq CC_1\varepsilon (s/t)^{2-3\delta} s^{-3/2}, && |I| + |J| \leq N - 8,\\
&\left|\partial^I L^J \partial_\alpha \partial_\beta \phi\right| \leq CC_1\varepsilon (s/t)^{1-3\delta} s^{-3/2+C(C_1\varepsilon)^{1/2}} && |I| + |J| \leq N - 8.
\end{aligned} \tag{11.9}$$

11.2 First refinement on the metric components

We begin the proof of the refined sup-norm estimate by the following bound on $L^J\left(h^{\mu\nu}\partial_\mu\partial_\nu h_{\alpha\beta}\right)$.

Lemma 11.1. *For all $|J| \leq N-7$, the following estimate holds:*

$$\left|L^J\left(h^{\mu\nu}\partial_\mu\partial_\nu h_{\alpha\beta}\right)\right| \leq C(C_1\varepsilon)^2 t^{-2+CC_1\varepsilon}(t-r)^{-1+CC_1\varepsilon}. \tag{11.10}$$

Proof. We have the following identity

$$\begin{aligned}h^{\mu\nu}\partial_\mu\partial_\nu h_{\alpha\beta} =& \underline{h}^{00}\partial_t\partial_t h_{\alpha\beta} + \underline{h}^{a0}\underline{\partial}_a\partial_t h_{\alpha\beta} + \underline{h}^{0b}\partial_t\underline{\partial}_b h_{\alpha\beta} + \underline{h}^{ab}\underline{\partial}_a\underline{\partial}_b h_{\alpha\beta} \\ &+ h^{\mu\nu}\partial_\mu\left(\Psi_\nu^{\nu'}\right)\underline{\partial}_{\nu'}h_{\alpha\beta}.\end{aligned}$$

We obtain

$$\begin{aligned}\left|L^J\left(h^{\mu\nu}\partial_\mu\partial_\nu h_{\alpha\beta}\right)\right| \leq& \left|L^J\left(\underline{h}^{00}\partial_t\partial_t h_{\alpha\beta}\right)\right| + \left|L^J\left(\underline{h}^{a0}\underline{\partial}_a\partial_t h_{\alpha\beta}\right)\right| \\ &+ \left|L^J\left(\underline{h}^{0b}\partial_t\underline{\partial}_b h_{\alpha\beta}\right)\right| + \left|L^J\left(\underline{h}^{ab}\underline{\partial}_a\underline{\partial}_b h_{\alpha\beta}\right)\right| \\ &+ \left|L^J\left(h^{\mu\nu}\partial_\mu\left(\Psi_\nu^{\nu'}\right)\underline{\partial}_{\nu'}h_{\alpha\beta}\right)\right|.\end{aligned}$$

The second, third, and fourth terms are null terms, they contain at least one "good" derivative and can be bounded directly by applying the basic sup-norm estimates. We only treat $\underline{h}^{a0}\underline{\partial}_a\partial_t h_{\alpha\beta}$, since the third and fourth terms are bounded similarly:

$$\left|L^J\left(\underline{h}^{a0}\underline{\partial}_a\partial_t h_{\alpha\beta}\right)\right| \leq \sum_{J_1+J_2=J}\left|L^{J_1}\underline{h}^{a0}L^{J_2}\underline{\partial}_a\partial_t h_{\alpha\beta}\right|.$$

We observe that

$$\left|L^{J_2}\underline{\partial}_a\partial_t h_{\alpha\beta}\right| = \left|L^{J_2}\left(t^{-1}L_a\partial_t h_{\alpha\beta}\right)\right| \leq \sum_{J_3+J_4=J_2}\left|L^{J_3}\left(t^{-1}\right)L^{J_4}L_a\partial_t h_{\alpha\beta}\right|.$$

Observe that $L^{J_3}\left(t^{-1}\right)$ is again smooth, homogenous of degree -1, which can be bounded by Ct^{-1} in $\mathcal{K}$. So the above sum is bounded by

$$\sum_{|J'|\leq|J|+1} Ct^{-1}\left|L^{J'}\partial_t h_{\alpha\beta}\right| \leq CC_1\varepsilon t^{-3/2}s^{-1+CC_1\varepsilon},$$

where we have applied (10.9). On the other hand, in view of (10.11), we have

$$\left|L^{J_1}\underline{h}^{a0}\right| \leq CC_1\varepsilon\left(t^{-1} + (s/t)t^{-1/2}s^{CC_1\varepsilon}\right),$$

since $\underline{h}^{a0}$ is a linear combination of $h_{\alpha\beta}$ with smooth and homogeneous coefficients of degree zero plus high order correction terms. We conclude that

$$\left|L^J\left(\underline{h}^{a0}\underline{\partial}_a\partial_t h_{\alpha\beta}\right)\right| \leq C(C_1\varepsilon)^2 t^{-3}s^{CC_1\varepsilon}.$$

Furthermore, the term $\left|L^J\left(h^{\mu\nu}\partial_\mu\left(\Psi_\nu^{\nu'}\right)\underline{\partial}_{\nu'}h_{\alpha\beta}\right)\right|$ is bounded by making use of the additional decay provided by $\left|L^{J'}\partial_\mu\left(\Psi_\nu^{\nu'}\right)\right| \leq C(J')t^{-1}$, and we omit the details and just state that

$$\left|L^J\left(h^{\mu\nu}\partial_\mu\left(\Psi_\nu^{\nu'}\right)\underline{\partial}_{\nu'}h_{\alpha\beta}\right)\right| \leq C(C_1\varepsilon)^2 t^{-3}s^{CC_1\varepsilon}.$$

Now we focus on the most problematic term $L^J\left(\underline{h}^{00}\partial_t\partial_t h_{\alpha\beta}\right)$. We apply here the sharp decay of $\underline{h}^{00}$ provided by (10.13) and the refined second-order estimate (10.15):

$$\begin{aligned}
&\left|L^J\left(\underline{h}^{00}\partial_t\partial_t h_{\alpha\beta}\right)\right| \leq \sum_{J_1+J_2=J}\left|L^{J_1}\underline{h}^{00}L^{J_2}\partial_t\partial_t h_{\alpha\beta}\right| \\
&\leq CC_1\varepsilon\left(t^{-1}+(s/t)^2t^{-1/2}s^{CC_1c}\right)CC_1\varepsilon t^{1/2}s^{-3+CC_1\varepsilon} \\
&\leq C(C_1\varepsilon)^2t^{-1/2}s^{-3+CC_1\varepsilon}+C(C_1\varepsilon)^2t^{-2}s^{-1+CC_1\varepsilon} \\
&\leq C(C_1\varepsilon)^2t^{-2+CC_1\varepsilon}(t-r)^{-1+CC_1\varepsilon}.
\end{aligned}$$

□

Lemma 11.2 (First refinement on $h_{\alpha\beta}$). *Assuming that the bootstrap assumption* (5.1) *holds with $C_1\varepsilon$ sufficiently small, one has*

$$|h_{\alpha\beta}| \leq CC_1\varepsilon t^{-1}s^{2\delta}. \tag{11.11}$$

Proof. We apply Proposition 3.4 and follow the notation therein. The wave equation satisfied by $h_{\alpha\beta}$

$$\tilde{\Box}_g h_{\alpha\beta} = F_{\alpha\beta} - 16\pi\phi\partial_\alpha\phi\partial_\beta\phi - 8\pi c^2\phi_{\alpha\beta}^2$$

leads us to

$$\Box h_{\alpha\beta} = -h^{\mu\nu}\partial_\mu\partial_\nu h_{\alpha\beta} + F_{\alpha\beta} - 16\pi\phi\partial_\alpha\phi\partial_\beta\phi - 8\pi c^2\phi^2 g_{\alpha\beta}.$$

We can apply (11.10) and (10.14), and we have

$$|S_{I,\alpha\beta}^W| \leq C(C_1\varepsilon)^2t^{-2+CC_1\varepsilon}(t-r)^{-1+CC_1\varepsilon}. \tag{11.12}$$

Second, by the basic sup-norm estimates, we have

$$|S_{\alpha\beta}^{KG,I,J}| \leq C(C_1\varepsilon)^2t^{-2-1/2+\delta}(t-r)^{-1/2+\delta}, \quad |I|+|J| \leq N-6.$$

We can choose $\varepsilon_2 > 0$ sufficiently small so that $\varepsilon \leq \varepsilon_2$ and $CC_1\varepsilon \leq \delta$, hence

$$|S_{I,\alpha\beta}^W[t,x,\partial^I L^J]| \leq C(C_1\varepsilon)^2t^{-2+\delta}(t-r)^{-1+\delta}$$

and, by Proposition 3.4,

$$|h_{\alpha\beta}(t,x)| \leq C(C_1\varepsilon)^2(t-r)^{2\delta}t^{-1} + CC_1\varepsilon t^{-1} \leq CC_1\varepsilon(t-r)^\delta t^{-1+\delta}. \quad \Box$$

11.3 First refinement for the scalar field

In this section, we apply Proposition 3.5 and consider first the correction terms.

Lemma 11.3. *Assume the bootstrap assumption* (5.1), (5.2) *and take the notation of Section 3.4 and Proposition 3.5, then for* $|I| + |J| \le N - 4$

$$|R_1[\partial^I L^J \phi]| \le CC_1\varepsilon (s/t)^{3/2} s^{-3/2+\delta}, \tag{11.13a}$$

$$|R_2[\partial^I L^J \phi]| \le C(C_1\varepsilon)^2 (s/t)^{3/2} s^{-3/2+3\delta}, \tag{11.13b}$$

$$|R_3[\partial^I L^J \phi]| \le C(C_1\varepsilon)^2 (s/t)^{3/2} s^{-3/2+3\delta}. \tag{11.13c}$$

Proof. We apply the basic sup-norm estimate to the corresponding expressions of R_i. For $R_1[\partial^I L^J \phi]$, we apply (4.20). For the term $R_2[\partial^I L^J \phi]$, we observe that $\left|\overline{h}^{00}\right| = \left|(t/s)^2 \underline{h}^{00}\right|$ and we recall that the linear part of $\underline{h}^{00}$ is a linear combination of $h_{\alpha\beta}$ with smooth and homogeneous coefficients of degree zero. We see that, in view of (11.11) (after neglecting the higher-order terms which vanish as $|h_{\alpha\beta}|^2$ at zero),

$$\left|\overline{h}^{00}\right| \le CC_1\varepsilon (s/t)^{-1} s^{-1+2\delta}.$$

Similarly, we have

$$\left|\overline{h}^{0b}\right| \le \left|(t/s)\underline{h}^{0b}\right|,$$

so that

$$\left|\overline{h}^{0b}\right| \le CC_1\varepsilon s^{-1+2\delta}$$

and, for $\overline{h}^{ab} = \underline{h}^{ab}$, we have $\left|\overline{h}^{ab}\right| \le CC_1\varepsilon (s/t)^2 s^{-1+2\delta}$. We also note that $\overline{\partial}_0\phi = (s/t)\partial_t\phi$. Then, substituting the above bounds leads us to

$$\left|R_2[\partial^I L^J \phi]\right| \le CC_1\varepsilon (s/t)^{3/2} s^{-3/2+3\delta}.$$

A similar derivation allows us to control

$$\left|R_3[\partial^I L^J \phi]\right| \le CC_1\varepsilon (s/t)^{3/2} s^{-3/2+3\delta}.$$

□

Proposition 11.1 (Estimate on ϕ and $\partial\phi$). *Assume the bootstrap assumption* (5.1) *and* (5.2) *hold with* $C_1 > C_0$ *and* $C_1\varepsilon$ *sufficiently small, then*

$$(s/t)^{3\delta-2}|\phi(t,x)| + (s/t)^{3\delta-3}|\underline{\partial}_\perp \phi(t,x)| \le CC_1\varepsilon s^{-3/2}. \tag{11.14}$$

Proof. We apply Proposition 3.5 and follow the notation there. Recall that Lemma 11.3 and Lemma 7.4, we have

$$
\begin{aligned}
|F(\tau)| &\leq \int_{s_0}^{\tau} \Big|\sum_i R_i[\phi](\lambda t/s, \lambda x/s)\Big| d\lambda \\
&\leq CC_1\varepsilon(s/t)^{3/2} \int_{s_0}^{\tau} \lambda^{-3/2+3\delta} d\lambda \leq CC_1\varepsilon(s/t)^{3/2} s_0^{-1/2+3\delta}
\end{aligned}
$$

and

$$|h'_{t,x}(\lambda)| \leq CC_1\varepsilon(s/t)^{1/2}\lambda^{-3/2+\delta} + CC_1\varepsilon(t/s)\lambda^{-2}.$$

We observe that, in the inequality (3.30) we need

$$
\begin{aligned}
\int_{\tau}^{s} |h'_{t,x}(\lambda) d\lambda| &\leq CC_1\varepsilon(s/t)^{1/2} \int_{s_0}^{s} \lambda^{-3/2+\delta} d\lambda + CC_1\varepsilon(s/t)^{-1} \int_{s_0}^{s} \lambda^{-2} d\lambda \\
&\leq CC_1\varepsilon(s/t)^{1/2} s_0^{-1/2+\delta} + CC_1\varepsilon(s/t)^{-1} s_0^{-1}.
\end{aligned}
$$

By (3.30), we have

$$|s^{3/2}\phi(t,x)| + \left|(s/t)^{-1}s^{3/2}\underline{\partial}_{\perp}\phi(t,x)\right| \leq V(t,x)$$

with

$$
V(t,x) \leq \begin{cases}
(\|v_0\|_{L^\infty} + \|v_1\|_{L^\infty})\left(1 + \int_2^s |h'_{t,x}(\overline{s})| e^{C\int_{\overline{s}}^{s}|h'_{t,x}(\lambda)|d\lambda}\right) & \\
+ F(s) + \int_2^s F(\overline{s})|h'_{t,x}(\overline{s})| e^{C\int_{\overline{s}}^{s}|h'_{t,x}(\lambda)|d\lambda} d\overline{s}, & 0 \leq r/t \leq 3/5, \\
F(s) + \int_{s_0}^{s} F(\overline{s})|h'_{t,x}(\overline{s})| e^{C\int_{\overline{s}}^{s}|h'_{t,x}(\lambda)d\lambda|} d\overline{s}, & 3/5 < r/t < 1.
\end{cases}
$$

When $0 \leq r/t \leq 3/5$, we get $4/5 \leq s/t \leq 1$ and $s_0 = 2$. This leads us to

$$V(t,x) \leq CC_1\varepsilon + CC_1\varepsilon \leq CC_1\varepsilon,$$

where we recall that $C_0 \leq C_1$. When $3/5 \leq r/t < 1$, the estimate is more delicate. In this case, we have $s_0 = \sqrt{\frac{t+r}{t-r}} \simeq (s/t)^{-1}$. This leads us to the following bounds:

$$|F(\tau)| \leq CC_1\varepsilon(s/t)^{2-3\delta}, \qquad \int_{\tau}^{s} |h'_{t,x}(\lambda) d\lambda| \leq CC_1\varepsilon.$$

Substituting these bounds into (3.30), we obtain

$$|s^{3/2}\phi(t,x)| + |(s/t)^{-1}s^{3/2}\underline{\partial}_{\perp}\phi(t,x)| \leq CC_1\varepsilon(s/t)^{2-3\delta}.$$

□

11.4 Second refinement for the scalar field and the metric

In this section, we establish the following result.

Lemma 11.4 (Second sup-norm refinement). *Assume that the bootstrap assumption* (5.1) *and* (5.2) *hold with $C_1 > C_0$ and $C_1\varepsilon$ sufficiently small, then for all $0 \leq |I| \leq N-7$,*

$$(s/t)^{3\delta-2}|\partial^I \phi| + (s/t)^{3\delta-3}|\underline{\partial}_\perp \partial^I \phi| \leq CC_1\varepsilon s^{-3/2}, \tag{11.15}$$

$$|h_{\alpha\beta}| \leq CC_1\varepsilon t^{-1} s^{C(C_1\varepsilon)^{1/2}}. \tag{11.16}$$

We need to control the commutators first.

Lemma 11.5. *For $|I| + |J| \leq N-7$,*

$$\begin{aligned}
&\big|[\partial^I L^J, h^{\mu\nu}\partial_\mu\partial_\nu]\phi\big| \leq C(C_1\varepsilon)^2 (s/t)^2 s^{-3+3\delta} \\
&+ \sum_{\substack{|J_1'|+|J_2'|\leq J \\ |J_2'|<|J|}} \left|L^{J_1'}\underline{h}^{00}\partial_t\partial_t\partial^I L^{J_2'}\phi\right| + \sum_{|J'|<|J|} \left|\underline{h}^{00}\partial_t\partial_t\partial^I L^{J'}\phi\right|.
\end{aligned} \tag{11.17}$$

Proof. We need to estimate all the terms listed in (4.16). As far as the terms $GQQ_{h\phi}$ are concerned, we will only treat in detail the term $\partial^{I_1}L^{J_1}h_{\alpha'\beta'}\partial^{I_2}L^{J_2}\underline{\partial}_a\partial_\mu\phi$. For $|I| + |J| \leq N-7$, we have

$$\begin{aligned}
&\left|\partial^{I_1}L^{J_1}h_{\alpha'\beta'}\partial^{I_2}L^{J_2}\underline{\partial}_a\partial_\mu\phi\right| \leq \left|\partial^{I_1}L^{J_1}h_{\alpha'\beta'}\right| \left|\partial^{I_2}L^{J_2}\underline{\partial}_a\partial_\mu\phi\right| \\
&\leq CC_1\varepsilon\left((s/t)t^{-1/2}s^\delta + t^{-1}\right)\left|\partial^{I_2}L^{J_2}\left(t^{-1}L_a\partial_\mu\phi\right)\right| \\
&\leq CC_1\varepsilon t^{-1}\left((s/t)t^{-1/2}s^\delta + t^{-1}\right) \sum_{\substack{|I_2'|\leq|I_2| \\ |J_2'|\leq|J_2|}} \left|\partial^{I_2'}L^{J_2'}L_a\partial_\mu\phi\right| \\
&\leq C(C_1\varepsilon)^2 t^{-3}s^{2\delta} = C(C_1\varepsilon)^2(s/t)^3 s^{-3+2\delta}.
\end{aligned}$$

Other terms of $GQQ_{h\phi}$ are bounded similarly, and we omit the details.

For the term $t^{-1}\partial^{I_3}L^{J_3}h_{\alpha'\beta'}\partial^{I_4}L^{J_4}\partial_\gamma\phi$, due to its additional t^{-1} decay, the basic sup-norm estimates are sufficient to get the following bound:

$$\begin{aligned}
\left|t^{-1}\partial^{I_3}L^{J_3}h_{\alpha'\beta'}\partial^{I_4}L^{J_4}\partial_\gamma\phi\right| &\leq C(C_1\varepsilon)^2 t^{-2}s^{-2+\delta} = C(C_1\varepsilon)^2(s/t)^2 s^{-4+2\delta} \\
&\leq C(C_1\varepsilon)^2(s/t)^3 s^{-3+2\delta}.
\end{aligned}$$

For the term $\partial^{I_1}L^{J_1}\underline{h}^{00}\partial^{I_2}L^{J_2}\partial_t\partial_t\phi$, we observe that $|I_1| \geq 1$, so it can be bounded in view of (7.1):

$$\begin{aligned}
\left|\partial^{I_1}L^{J_1}\underline{h}^{00}\partial^{I_2}L^{J_2}\partial_t\partial_t\phi\right| &\leq C(C_1\varepsilon)^2 t^{-3/2}s^\delta\, t^{-1/2}s^{-1+\delta} \\
&\leq C(C_1\varepsilon)^2(s/t)^2 s^{-3+2\delta}.
\end{aligned}$$

For the remaining terms in (4.16) we observe that the term $\partial^I L^{J_2'}\partial_t\partial_t\phi$ and $\partial_\gamma\partial_{\gamma'}\partial^I L^{J'}\phi$ are bounded by $\partial_t\partial_t\partial^{I'} L^{J'}\phi$ plus some corrections: $\left|\partial^I L^{J_2'}\partial_t\partial_t\phi\right| \leq C\sum_{\substack{\gamma,\gamma'\\|J_2''|\leq|J_2'|}}\left|\partial_\gamma\partial_{\gamma'}\partial^I L^{J_2''}\phi\right|$. Then in view of (7.23) and the argument presented below it (but now ϕ plays the role of $h_{\alpha\beta}$ in (7.23)), we have

$$\left|\partial^I L^{J_2'}\partial_t\partial_t\phi\right| \leq CC_1\varepsilon t^{-5/2}s^\delta + C\sum_{|J_2''|\leq|J_2'|}\left|\partial_t\partial_t\partial^I L^{J_2''}\phi\right|.$$

So the last two terms in (4.16) is bounded by

$$C(C_1\varepsilon)^2t^{-3}s^{2\delta}+C\sum_{\substack{|J_1'|+|J_2'|\leq|I|\\|J'|<|J|}}|L^{J_1'}\underline{h}^{00}\partial_t\partial_t\partial^I L^{J_2'}\phi|+C\sum_{|J'|<|J|}|\underline{h}^{00}\partial_t\partial_t\partial^I L^{J'}\phi|.$$

This yields us the conclusion. On the other hand, when $|J|=0$, the last two terms do not exist. □

Proof of Lemma 11.4. The proof of (11.15) is similar to that of Proposition 11.1. The only difference is that we need to bound the commutator $[\partial^I, h^{\mu\nu}\partial_\mu\partial_\nu]\phi$ (which, with the notation in Proposition 3.5, plays the role of f in the definition of F). We apply (11.17) with $|J|=0$ and, in this case, $\left|[\partial^I, h^{\mu\nu}\partial_\mu\partial_\nu]\phi\right| \leq C(C_1\varepsilon)^2(s/t)^2s^{-3+3\delta}$.

Then (following the notation in Proposition 3.5) in view of (11.2) and by an argument similar to the one in the proof of Proposition 11.1, we have

$$\begin{aligned}
|F(\tau)| &\leq CC_1\varepsilon(s/t)^{3/2}s_0^{-1/2+3\delta} + C(C_1\varepsilon)^2(s/t)^2s_0^{-1/2+3\delta},\\
|h_{t,x}'(\lambda)| &\leq CC_1\varepsilon(s/t)^{1/2}\lambda^{-3/2+\delta} + CC_1\varepsilon(t/s)\lambda^{-2},\\
\int_\tau^s |h_{t,x}'(\lambda)d\lambda| &\leq CC_1\varepsilon(s/t)^{1/2}s_0^{-1/2+\delta} + CC_1\varepsilon(s/t)^{-1}s_0^{-1}.
\end{aligned}$$

In view of (3.30), the desired results are thus proven.

The proof of (11.16) is an application of (11.15). We rely on the proof of Lemma 11.2 and we have that (11.12) still holds. We furthermore observe that in view of (11.15),

$$|S_{\alpha\beta}^{KG,I,J}| \leq C(C_1\varepsilon)^2t^{-3},\quad |I|+|J|\leq N-7.$$

Furthermore, since $C_1\varepsilon\leq 1$, we take, in view of (11.12)

$$\begin{aligned}
|S_{I,\alpha\beta}^W| &\leq C(C_1\varepsilon)^2t^{-2+CC_1\varepsilon}(t-r)^{-1+CC_1\varepsilon}\\
&\leq C(C_1\varepsilon)^2t^{-2+C(C_1\varepsilon)^{1/2}}(t-r)^{-1+C(C_1\varepsilon)^{1/2}}.
\end{aligned}$$

In view of Proposition 3.4, we arrive at

$$\begin{aligned}
|h_{\alpha\beta}| &\leq CC_1\varepsilon t^{-1} + \frac{C(C_1\varepsilon)^2}{CC_1\varepsilon}t^{-1+C(C_1\varepsilon)^{1/2}}(t-r)^{C(C_1\varepsilon)^{1/2}}\\
&\leq C(C_1\varepsilon)t^{-1}s^{C(C_1\varepsilon)^{1/2}}.\quad □
\end{aligned}$$

11.5 A secondary bootstrap argument

In this section, we improve the L^∞ bounds of $\partial^I L^J \phi$ and $\underline{\partial}_\perp \partial^I L^J \phi$ for $|I| + |J| \leq N - 7$.

Proposition 11.2. *There exists a pair of positive constants (C_1, ε_2) with $C_1 > C_0$ such that if (5.1) and (5.2) hold with C_1 and $0 \leq \varepsilon \leq \varepsilon_2$, then for all $|I| + |J| \leq N - 7$,*

$$(s/t)^{3\delta-2}|\partial^I L^J \phi| + (s/t)^{3\delta-3}|\underline{\partial}_\perp \partial^I L^J \phi| \leq CC_1\varepsilon s^{-3/2+C(C_1\varepsilon)^{1/2}}, \quad (11.18)$$

$$|L^J h_{\alpha\beta}| \leq CC_1\varepsilon t^{-1} s^{C(C_1\varepsilon)^{1/2}}. \quad (11.19)$$

Proof. We proceed by induction, by relying on a secondary bootstrap argument. Recall that the bootstrap assumptions (5.1) and (5.2) hold on $[2, s^*]$, and we suppose that there exist constants $K_{m-1}, C_{m-1} > 0$ and $\varepsilon'_{m-1} > 0$ depending only on the structure of the main system such that

$$(s/t)^{3\delta-2}|\partial^I L^J \phi| + (s/t)^{3\delta-3}|\underline{\partial}_\perp \partial^I L^J \phi| \leq K_{m-1}C_1\varepsilon s^{-3/2+C_{m-1}(C_1\varepsilon)^{1/2}}, \quad (11.20)$$

$$|L^J h_{\alpha\beta}(t,x)| \leq K_{m-1}C_1\varepsilon t^{-1} s^{C_{m-1}(C_1\varepsilon)^{1/2}} \quad (11.21)$$

holds on $[2, s^*]$ for all $0 \leq \varepsilon \leq \varepsilon'_{m-1}$ and $|J| \leq m - 1 \leq N - 7$ and $|I| + |J| \leq N - 7$. This is true when $|J| = 0$, guaranteed in view of (11.15) and (11.16) (since there the constant C depends only on N and the structure of the main system). We want prove that there exist constants K_m, C_m, ε'_m depending only on the structure of the main system such that

$$(s/t)^{3\delta-2}|\partial^I L^J \phi| + (s/t)^{3\delta-3}|\underline{\partial}_\perp \partial^I L^J \phi| \leq K_m C_1\varepsilon s^{-3/2+C_m(C_1\varepsilon)^{1/2}}, \quad (11.22)$$

$$|L^J h_{\alpha\beta}(t,x)| \leq K_m C_1\varepsilon t^{-1} s^{C_m(C_1\varepsilon)^{1/2}} \quad (11.23)$$

hold for $0 \leq \varepsilon \leq \varepsilon'_m$ and all $|J| \leq N - 7$.

We observe that on the initial slice $\mathcal{H}_2 \cap \mathcal{K}$, there exits a positive constant $K_{0,m}$ such that

$$(s/t)^{3\delta-2}|\partial^I L^J \phi| + (s/t)^{3\delta-3}|\underline{\partial}_\perp \partial^I L^J \phi| \leq K_{0,m}C_0\,\varepsilon \leq K_{0,m}C_1\varepsilon.$$

We also denote by $K_{0,m}$ a positive constant such that

$$\sup_{t=2,|x|\leq 1}\{ts^{-C_m(C_1\varepsilon)^{1/2}}|L^J h_{\alpha\beta}(t,x)|\} \leq K_{0,m}C_0\,\varepsilon \leq K_{0,m}C_1\varepsilon,$$

since we have chosen $C_1 \geq C_0$. Here we observe that on $\{t=2\} \cap \mathcal{K}$, $\sqrt{3} \leq s \leq 2$, so when $C_m > 0$, the constant $K_{0,m}$ can be chosen independently of C_m.

So, first, we choose $K_m > K_{0,m}$ and set

$$s^{**} := \sup_{s\in[2,s^*]} \{(11.22) \text{ and } (11.23) \text{ holds in } \mathcal{K}_{[2,s^{**}]}\}.$$

By continuity ($K_m > K_{0,m}$) we obtain $s^{**} > 2$. We prove that if we choose ε'_m sufficiently small, then for all $\varepsilon \leq \varepsilon'_m$, $s^{**} = s^*$. This is done as follows.

We take $K_m \geq K_{m-1}$, $C_m = 2C_{m-1}$ and see first that under the induction assumptions (11.20), (11.21) and the bootstrap assumptions (11.22) and (11.23), (11.17) becomes (in $\mathcal{K}_{[2,s^{**}]}$)

$$\begin{aligned}
&\left|[\partial^I L^J, h^{\mu\nu}\partial_\mu\partial_\nu]\phi\right| \\
&\leq C(C_1\varepsilon)^2(s/t)^2 s^{-3+3\delta} + CK_m^2(C_1\varepsilon)^2(s/t)^{2-3\delta}s^{-5/2+C_m(C_1\varepsilon)^{1/2}}.
\end{aligned}$$

We observe that, in the right-hand side of (11.17), the last term is bounded directly by applying (11.16) and (11.23). The second term is more delicate. We distinguish between two different cases. When $|J'_2| = 0$, we apply the bootstrap assumptions (11.23) and (11.15). When $0 < |J'_2| < |J|$, we have $|J'_1| \leq m-1$, so we apply (11.20) and (11.21) and observe that we have chosen $C_m = 2C_{m-1}$.

We then recall Lemma 11.3 and, by Proposition 3.5 (following the notation therein), we have in both cases $0 \leq r/t \leq 3/5$ and $3/5 < r/t < 1$,

$$\begin{aligned}
|F(s)| &\leq CC_1\varepsilon(s/t)^{3/2}\int_{s_0}^{s}\tau^{-3/2+3\delta}d\tau + CK_m^2(C_1\varepsilon)^2\int_{s_0}^{s}\tau^{-1+C_m(C_1\varepsilon)^{1/2}}d\tau \\
&\leq CC_1\varepsilon(s/t)^{3/2}s_0^{-1/2+3\delta} + CC_m^{-1}K_m^2(C_1\varepsilon)^{3/2}(s/t)^{2-3\delta}s^{C_m(C_1\varepsilon)^{1/2}} \\
&\leq CC_1\varepsilon(s/t)^{2-3\delta} + CC_m^{-1}K_m^2(C_1\varepsilon)^{3/2}(s/t)^{2-3\delta}s^{C_m(C_1\varepsilon)^{1/2}}.
\end{aligned}$$

We also have, in view of (7.10), $|h_{t,x}(\lambda)| \leq CC_1\varepsilon(s/t)^{1/2}\lambda^{-3/2+\delta} + CC_1\varepsilon(s/t)^{-1}\lambda^{-2}$ and then, in both cases $0 \leq r/t \leq 3/5$ and $3/5 < r/t < 1$,

$$\begin{aligned}
\int_{s_0}^{s}|h_{t,x}(\lambda)| &\leq CC_1\varepsilon(s/t)^{1/2}\int_{s_0}^{s}\lambda^{-3/2+\delta}d\lambda + CC_1\varepsilon(s/t)^{-1}\int_{s_0}^{s}\lambda^{-2}d\lambda \\
&\leq CC_1\varepsilon\left((s/t)^{1/2}s_0^{-1+\delta} + (s/t)^{-1}s_0^{-1}\right) \leq CC_1\varepsilon.
\end{aligned}$$

By Proposition 3.5, we have

$$\begin{aligned}
&(s/t)^{3\delta-2}s^{-3/2}\left|\partial^I L^J\phi\right| + (s/t)^{3\delta-3}s^{-3/2}\left|\underline{\partial}_\perp\partial^I L^J\phi\right| \\
&\leq CK_{0,m}C_1\varepsilon + CC_1\varepsilon + CC_m^{-1}K_m^2(C_1\varepsilon)^{3/2}s^{C_m(C_1\varepsilon)^{1/2}}.
\end{aligned}$$

We can choose K_m sufficiently large and fix $\varepsilon'_m = \frac{C_m^2}{C_1}\left(\frac{K_m-2CK_{0,m}-2C}{2CK_m^2}\right)^2 > 0$, and then we see that on $[2, s^{**}]$:

$$(s/t)^{3\delta-2}s^{-3/2}\,|\partial^I L^J\phi| + (s/t)^{3\delta-3}s^{-3/2}\,|\underline{\partial}_\perp\partial^I L^J\phi| \le \frac{1}{2}K_m C_1\varepsilon s^{C_m(C_1\varepsilon)^{1/2}}. \tag{11.24}$$

Here we need to emphasize that C_m is determined only by N and the structure of the system: we have C_0, determined in view of (11.16) where the constant C is determined by N and the main system. Then, $C_m = 2C_{m-1}$ thus C_m are determined only by N and the structure of the system.

In the same way, we follow the notation in Proposition 3.4 combined with following estimates deduced from (11.22): as $|I|+|J|\le N-7$

$$\begin{aligned}|S^{KG,I,J}_{\alpha\beta}| &\le C_m(C_1\varepsilon)^2(s/t)^{4-6\delta}s^{-3+C_m(C_1\varepsilon)^{1/2}}\\ &\le C(K_mC_1\varepsilon)^2t^{-3+3\delta+\frac{1}{2}C_m(C_1\varepsilon)^{1/2}}(t-r)^{-3\delta+\frac{1}{2}C_m(C_1\varepsilon)^{1/2}},\end{aligned}$$

where we rely on a similar argument for the estimate of $\big|[\partial^I L^J, h^{\mu\nu}\partial_\mu\partial_\nu]\phi\big|$.

We also recall (11.12) for $|I|+|J|\le N-7$

$$\begin{aligned}|S^W_{I,\alpha\beta}| &\le C(C_1\varepsilon)^2t^{-2+CC_1\varepsilon}(t-r)^{-1+CC_1\varepsilon}\\ &\le C(C_1\varepsilon)^2t^{-2+C(C_1\varepsilon)^{1/2}}(t-r)^{-1+C(C_1\varepsilon)^{1/2}}.\end{aligned}$$

This leads us to (by Proposition 3.4)

$$\begin{aligned}&|\partial^I L^J h_{\alpha\beta}|\\ &\le Cm_S\varepsilon t^{-1} + \frac{C(C_1\varepsilon)^2}{CC_1\varepsilon}t^{-1+C(C_1\varepsilon)^{1/2}}(t-r)^{C(C_1\varepsilon)^{1/2}}\\ &\quad + C(K_mC_1\varepsilon)^2t^{-1}s^{C_m(C_1\varepsilon)^{1/2}}\\ &\le CC_1K_{0,m}\varepsilon t^{-1} + CC_1\varepsilon t^{-1+C(C_1\varepsilon)^{1/2}}(t-r)^{C(C_1\varepsilon)^{1/2}}\\ &\quad + C(K_mC_1\varepsilon)^2t^{-1}(t-r)^{C_m(C_1\varepsilon)^{1/2}}\\ &\le CC_1\varepsilon\left(K_{0,m}+1+K_m^2C_1\varepsilon\right)t{-1} + C_m(C_1\varepsilon)^{1/2}(t-r)^{C_m(C_1\varepsilon)^{1/2}}.\end{aligned}$$

We check that when $\varepsilon\le\varepsilon'_m$, on $[2, s^{**}]$:

$$|\partial^I L^J h_{\alpha\beta}| \le \frac{1}{2}K_mC_1\varepsilon. \tag{11.25}$$

Now, in view of (11.24) and (11.25), we make the following observation: when $s^{**} < s^*$, by continuity we must have

$$(s/t)^{3\delta-2}|\partial^I L^J\phi| + (s/t)^{3\delta-3}|\underline{\partial}_\perp\partial^I L^J\phi| = K_mC_1\varepsilon s^{-3/2+C(C_1\varepsilon)^{1/2}} \tag{11.26}$$

or

$$|L^J h_{\alpha\beta}(t,x)| = K_mC_1\varepsilon t^{-1}s^{C(C_1\varepsilon)^{1/2}}. \tag{11.27}$$

This is a contradiction with (11.24) together with (11.25). We conclude that $s^{**} = s^*$. That is, (11.18) and (11.19) are proved for $|J| = m$. By induction, (11.18) and (11.19) are proved for $|J| \leq N - 7$. This concludes the argument, by taking $\varepsilon_2 = \varepsilon'_{N-7}$. □

Chapter 12

High-Order Refined L^2 Estimates

12.1 Objective of this section and preliminary

In this section we improve the energy bounds of both $h_{\alpha\beta}$ and ϕ for $N-4 \leq |I|+|J| \leq N$. We rely on the energy estimates Proposition 3.1 and Proposition 3.2. In order to apply these two propositions, we need a control of the source terms:

- For $\partial^I L^J h_{\alpha\beta}$, we have the terms $\partial^I L^J F_{\alpha\beta}$, QS_ϕ, $[\partial^I L^J, h^{\mu\nu}\partial_\mu\partial_\nu]h_{\alpha\beta}$.
- For $\partial^I L^J \phi$, we have the terms $[\partial^I L^J, h^{\mu\nu}\partial_\mu\partial_\nu]\phi$.

In this section, we derive the L^2 bounds and apply them (in the next subsection) in the proof of the main estimate. Note that the estimate for $F_{\alpha\beta}$ is already covered by Lemma 10.1. We begin with QS_ϕ.

Lemma 12.1. *Assume the bootstrap assumptions* (5.1) *and* (5.2) *hold. Then the following estimates hold for* $|I|+|J| \leq N$*:*

$$\begin{aligned}
&\left\|\partial^I L^J\left(\partial_\alpha\phi\partial_\beta\phi\right)\right\|_{L^2(\mathcal{H}^*_s)} + \left\|\partial^I L^J\left(\phi^2\right)\right\|_{L^2(\mathcal{H}^*_s)} \\
&\leq CC_1\varepsilon s^{-3/2}\sum_{|I'|\leq|I|} E_{M,c^2}(s,\partial^{I'}L^J\phi)^{1/2} \\
&\quad + CC_1\varepsilon s^{-3/2+C(C_1\varepsilon)^{1/2}} \sum_{\substack{|I'|\leq|I|\\|J'|<|J|}} E_{M,c^2}(s,\partial^{I'}L^{J'}\phi)^{1/2}.
\end{aligned} \tag{12.1}$$

Proof. We only treat $\partial^I L^J(\partial_\alpha\phi\partial_\beta\phi)$ and omit the argument for $\partial^I L^J(\phi^2)$ which is simpler. We have $\partial^I L^J(\partial_\alpha\phi\partial_\beta\phi) = \sum_{\substack{I_1+I_2=I\\J_1+J_2=J}} \partial^{I_1}L^{J_1}\partial_\alpha\phi\partial^{I_2}L^{J_2}\partial_\beta\phi$. Assuming that $N \geq 13$, we have either $|I_1|+|J_1| \leq N-7$ or $|I_2|+|J_2| \leq N-7$. Without loss of generality, we suppose that $|I_1|+|J_1| \leq N-7$:

- When $|I_1| = |J_1| = 0$. We apply (11.8):
$$\|\partial^{I_1} L^{J_1} \partial_\alpha \phi \, \partial^{I_2} L^{J_2} \partial_\beta \phi\|_{L^2(\mathcal{H}_s^*)} = \|\partial_\alpha \phi \, \partial^I L^J \partial_\beta \phi\|_{L^2(\mathcal{H}_s^*)}$$
$$\le CC_1\varepsilon \left\|(s/t)^{2-3\delta} s^{-3/2} (t/s)\, (s/t) \partial^I L^J \partial_\beta \phi\right\|_{L^2(\mathcal{H}_s^*)}$$
$$\le CC_1 \varepsilon s^{-3/2} E_{M,c^2}(s, \partial^I L^J \phi)^{1/2}.$$
- When $|J_1| = 0, 1 \le |I_1| \le N-7$, then $|I_2| + |J_2| \le N-1$. We apply (11.6):
$$\|\partial^{I_1} L^{J_1} \partial_\alpha \phi \, \partial^{I_2} L^{J_2} \partial_\beta \phi\|_{L^2(\mathcal{H}_s^*)} = \|\partial^{I_1} \partial_\alpha \phi \, \partial^{I_2} L^J \partial_\beta \phi\|_{L^2(\mathcal{H}_s^*)}$$
$$\le CC_1\varepsilon \left\|(s/t)^{1-3\delta} s^{-3/2} \, \partial^{I_2} L^J \partial_\beta \phi\right\|_{L^2(\mathcal{H}_s^*)}$$
$$\le CC_1 \varepsilon s^{-3/2} \sum_{|I'| \le |I|} E_{M,c^2}(\partial^{I'} L^J \phi)^{1/2}.$$
- When $1 \le |J_1|$ and $|I_1| + |J_1| \le N-7$, then $|I_2| + |J_2| \le N-1$ and $|J_2| < |J|$. We apply (11.4)
$$\|\partial^{I_1} L^{J_1} \partial_\alpha \phi \, \partial^{I_2} L^{J_2} \partial_\beta \phi\|_{L^2(\mathcal{H}_s^*)}$$
$$\le CC_1\varepsilon \left\|(s/t)^{1-3\delta} s^{-3/2+C(C_1\varepsilon)^{1/2}} \, \partial^{I_2} L^{J_2} \partial_\beta \phi\right\|_{L^2(\mathcal{H}_s^*)}$$
$$\le CC_1 \varepsilon s^{-1+C(C_1\varepsilon)^{1/2}} s^{-1/2} \sum_{\substack{I' \le |I| \\ |J'| < |J|}} E_{M,c^2}(s, \partial^{I'} L^{J'} \phi)^{1/2}. \quad \square$$

Lemma 12.2. *Under the bootstrap assumption, for $|I| + |J| \le N$ one has*
$$\begin{aligned}
&\|[\partial^I L^J, h^{\mu\nu} \partial_\mu \partial_\nu] h_{\alpha\beta}\|_{L^2(\mathcal{H}_s^*)} \\
&\le CC_1 \varepsilon s^{-1} \sum_{\substack{\alpha', \beta', a, |I'| \le |I| \\ |J'| < |J|}} E_M^*(s, \partial^{I'} L_a L^{J'} h_{\alpha'\beta'})^{1/2} \\
&\quad + CC_1 \varepsilon s^{-1+C(C_1\varepsilon)} \sum_{\substack{\alpha' \beta', |I'| \le |I| \\ |J'| < |J|}} E_M^*(s, \partial^{I'} L^{J'} h_{\alpha'\beta'})^{1/2} \\
&\quad + CC_1 \varepsilon s^{-3/2} \sum_{|I'| \le |I|} E_{M,c^2}^*(s, \partial^{I'} L^J \phi)^{1/2} \\
&\quad + CC_1 \varepsilon s^{-3/2+C(C_1\varepsilon)^{1/2}} \sum_{\substack{|I'| \le |I| \\ |J'| < |J|}} E_{M,c^2}^*(s, \partial^{I'} L^{J'} \phi)^{1/2} + C(C_1\varepsilon)^2 s^{-3/2+3\delta}
\end{aligned} \tag{12.2}$$
and, in particular, for $|J| = 0$,
$$\|[\partial^I, h^{\mu\nu} \partial_\mu \partial_\nu] h_{\alpha\beta}\|_{L^2(\mathcal{H}_s^*)}$$
$$\le CC_1 \varepsilon s^{-3/2} \sum_{|I'| \le |I|} E_{M,c^2}^*(s, \partial^{I'} \phi)^{1/2} + C(C_1\varepsilon)^2 s^{-3/2+3\delta}.$$

Proof. We rely on the estimate (8.20) and (8.5) combined with (12.1). In view of (8.20), we need to estimate $\left\|(s/t)^2\partial_t\partial_t\partial^I L^{J'} h_{\alpha\beta}\right\|_{L^2(\mathcal{H}_s^*)}$ for $|J'| < |J|$. Then, in view of (8.5), the above quantity is to be bounded by the L^2 norm of $Sc_1[\partial^I L^{J'} h_{\alpha\beta}]$, $Sc_2[\partial^I L^{J'} h_{\alpha\beta}]$, $\partial^I L^{J'} F_{\alpha\beta}$, and $\partial^I L^{J'} QS_\phi$. These terms are bounded respectively in view of (8.10), (8.12), Lemma 10.1 and (12.1). With all these estimate substitute into (8.5), we have for $|J'| < |J|$,

$$
\begin{aligned}
&\left\|(s/t)^2\partial_t\partial_t\partial^I L^{J'} h_{\alpha\beta}\right\|_{L^2(\mathcal{H}_s^*)} \\
&\leq Cs^{-1} \sum_{\substack{\alpha',\beta',a,|I'|\leq|I| \\ |J'|<|J|}} E_M^*(s, \partial^{I'} L_a L^{J'} h_{\alpha'\beta'})^{1/2} \\
&\quad + CC_1\varepsilon s^{-1+C(C_1\varepsilon)} \sum_{\substack{\alpha'\beta',|I'|\leq|I| \\ |J'|<|J|}} E_M^*(s, \partial^{I'} L^{J'} h_{\alpha'\beta'})^{1/2} \\
&\quad + CC_1\varepsilon s^{-3/2} \sum_{|I'|\leq|I|} E_{M,c^2}^*(s, \partial^{I'} L^J \phi)^{1/2} \\
&\quad + CC_1\varepsilon s^{-3/2+C(C_1\varepsilon)^{1/2}} \sum_{\substack{|I'|\leq|I| \\ |J'|<|J|}} E_{M,c^2}^*(s, \partial^{I'} L^{J'} \phi)^{1/2} \\
&\quad + \sum_{|J'|<|J|} \|[\partial^I L^{J'}, h^{\mu\nu}\partial_\mu\partial_\nu]h_{\alpha\beta}\|_{L_f^2(\mathcal{H}_s)} + C(C_1\varepsilon)^2 s^{-3/2+2\delta}.
\end{aligned}
\tag{12.3}
$$

That is, we have

$$
\begin{aligned}
&\|[\partial^I L^J, h^{\mu\nu}\partial_\mu\partial_\nu]h_{\alpha\beta}\|_{L^2(\mathcal{H}_s^*)} \\
&\leq CC_1\varepsilon s^{-1} \sum_{\substack{\alpha',\beta',a,|I'|\leq|I| \\ |J'|<|J|}} E_M^*(s, \partial^{I'} L_a L^{J'} h_{\alpha'\beta'})^{1/2} \\
&\quad + CC_1\varepsilon s^{-1+C(C_1\varepsilon)} \sum_{\substack{\alpha'\beta',|I'|\leq|I| \\ |J'|<|J|}} E_M^*(s, \partial^{I'} L^{J'} h_{\alpha'\beta'})^{1/2} \\
&\quad + CC_1\varepsilon s^{-3/2} \sum_{|I'|\leq|I|} E_{M,c^2}^*(s, \partial^{I'} L^J \phi)^{1/2} \\
&\quad + CC_1\varepsilon s^{-3/2+C(C_1\varepsilon)^{1/2}} \sum_{\substack{|I'|\leq|I| \\ |J'|<|J|}} E_{M,c^2}^*(s, \partial^{I'} L^{J'} \phi)^{1/2} \\
&\quad + \sum_{|J'|<|J|} \|[\partial^I L^{J'}, h^{\mu\nu}\partial_\mu\partial_\nu]h_{\alpha\beta}\|_{L_f^2(\mathcal{H}_s)} + C(C_1\varepsilon)^2 s^{-3/2+2\delta}.
\end{aligned}
$$

Then, we proceed by induction on J and the desired result is reached. When $|J| = 0$, in the right-hand side of the above estimate there exist only the third and the last term, this proves the desired result in this case. Then, by induction on $|J|$, the desired result is established for $|I| + |J| \leq N$. □

Lemma 12.3. *Under the bootstrap assumption, for all $|I| + |J| \leq N$ one has*

$$
\begin{aligned}
&\big\|[\partial^I L^J, h^{\mu\nu}\partial_\mu\partial_\nu]\phi\big\|_{L^2_f(\mathcal{H}_s)} \\
&\leq CC_1\varepsilon s^{-1/2} \sum_{\substack{|J'|=|J| \\ \alpha,\beta}} E^*_M(s, L^{J'} h_{\alpha\beta})^{1/2} \\
&\quad + CC_1\varepsilon s^{-1/2} \sum_{\substack{|J'|=|J| \\ \alpha\beta}} \int_2^s \tau^{-1} E^*_M(\tau, L^{J'} h_{\alpha\beta})^{1/2} d\tau \\
&\quad + CC_1\varepsilon s^{-1+C(C_1\varepsilon)^{1/2}} \sum_{\substack{|I'|\leq|I|+1 \\ |J'|<|J|}} E^*_M(s, \partial^{I'} L^{J'} \phi)^{1/2} \\
&\quad + CC_1\varepsilon s^{-1/2+C(C_1\varepsilon)^{1/2}} \sum_{\substack{|J_1'|<|J| \\ \alpha',\beta'}} E^*_M(s, L^{J_1'} h_{\alpha'\beta'})^{1/2} \\
&\quad + CC_1\varepsilon s^{-1/2+C(C_1\varepsilon)^{1/2}} \sum_{\substack{|J_1'|<|J| \\ \alpha',\beta'}} \int_2^s \tau^{-1} E^*_M(\tau, L^{J_1'} h_{\alpha'\beta'})^{1/2} d\tau \\
&\quad + C(C_1\varepsilon)^2 s^{-1/2+C(C_1\varepsilon)^{1/2}}.
\end{aligned}
\tag{12.4}
$$

When $|J| = 0$, one has

$$\big\|[\partial^I, h^{\mu\nu}\partial_\mu\partial_\nu]\phi\big\|_{L^2_f(\mathcal{H}_s)} \leq C(C_1\varepsilon)^2 s^{-1+3\delta}. \tag{12.5}$$

Proof. We need to estimate the terms listed in (4.16). The estimates on first two terms are trivial: one is a null term and the other has a additional decay t^{-1}. We just point out that for the first term we need to apply (4.18), (4.19) combined with (5.22) or (3.39) and write down their L^2 bounds

$$\|\partial^I L^J GQQ_{h\phi}\|_{L^2(\mathcal{H}^*_s)} + \|t^{-1}\partial^{I_1} L^{J_1} h_{\mu\nu} \partial^{I_2} L^{J_2} \partial_\gamma \phi\|_{L^2(\mathcal{H}^*_s)} \leq C(C_1\varepsilon)^2 s^{-1+2\delta}. \tag{12.6}$$

We focus on the last three terms.

Term 1. $\partial^{I_1} L^{J_1} \underline{h}^{00} \partial^{I_2} L^{J_2} \partial_t\partial_t\phi$. Recall that $|I_1| \geq 1$. The L^2 norm of this term is bounded by a discussion on the following cases:

- Case $1 \leq |I_1| + |J_1| \leq N - 2$. We apply (7.1) combined with the basic energy estimate:

$$\begin{aligned}
&\left\|\partial^{I_1}L^{J_1}\underline{h}^{00}\partial^{I_2}L^{J_2}\partial_t\partial_t\phi\right\|_{L^2(\mathcal{H}^*_s)}\\
&\leq CC_1\varepsilon\left\|t^{-3/2}s^{\delta}(t/s)\,(s/t)\partial^{I_2}L^{J_2}\partial_t\partial_t\phi\right\|_{L^2(\mathcal{H}^*_s)}\\
&\leq C(C_1\varepsilon)^2s^{-1+3\delta}.
\end{aligned}$$

• Case $N-1\leq|I_1|+|J_1|\leq N$, then $|I_2|+|J_2|\leq 1\leq N-8$. Then we apply (7.12) combined with the basic sup-norm estimate for $\partial^{I_2}L^{J_2}\partial_t\partial_t\phi$:

$$\begin{aligned}
\left\|\partial^{I_1}L^{J_1}\underline{h}^{00}\partial^{I_2}L^{J_2}\partial_t\partial_t\phi\right\|_{L^2(\mathcal{H}^*_s)}&\leq CC_1\varepsilon\left\|(s/t)\partial^{I_1}L^{J_1}\underline{h}^{00}\,(t/s)t^{-3/2}s^{\delta}\right\|_{L^2(\mathcal{H}^*_s)}\\
&\leq CC_1\varepsilon s^{-3/2+\delta}\left\|(s/t)\partial^{I_1}L^{J_1}\underline{h}^{00}\right\|_{L^2(\mathcal{H}^*_s)}\\
&\leq C(C_1\varepsilon)^2s^{-3/2+3\delta}.
\end{aligned}$$

Term 2. $L^{J_1}\underline{h}^{00}\partial^{I}L^{J_2}\partial_t\partial_t\phi$. Recall that $|J_1|\geq 1$ so that $|J_2|\leq|J|-1\leq N-1$.

• Case $1\leq|J_1|\leq N-7$. In this case, we apply (11.19) to $L^{J_1}\underline{h}^{00}$ (seen as a linear combination of $L^{J_1'}h_{\alpha\beta}$ with $|J_1'|$ plus higher-order corrections):

$$\begin{aligned}
\left\|L^{J_1}\underline{h}^{00}\partial^{I}L^{J_2}\partial_t\partial_t\phi\right\|_{L^2(\mathcal{H}^*_s)}&\leq CC_1\varepsilon\left\|t^{-1}s^{C(C_1\varepsilon)^{1/2}}\partial^{I}L^{J_2}\partial_t\partial_t\right\|_{L^2(\mathcal{H}^*_s)}\\
&\leq CC_1\varepsilon s^{-1+C(C_1\varepsilon)^{1/2}}\left\|(s/t)\partial^{I}L^{J_2}\partial_t\partial_t\right\|_{L^2(\mathcal{H}^*_s)}\\
&\leq CC_1\varepsilon s^{-1+C(C_1\varepsilon)^{1/2}}\sum_{|J'|<|J|}E_{M,c^2}(s,\partial^{I}L^{J'}\phi)^{1/2}.
\end{aligned}$$

• Case $N-6\leq|J_1|\leq|J|-1\leq N-1$ then $|I|+|J_2|\leq 6\leq N-8$. In this case we apply Proposition 7.1 to $L^{J_1}\underline{h}^{00}$ and (11.4). First of all, by the estimates (3.54) of commutators and (11.4), we deduce that $\left|\partial^{I}L^{J_2}\partial_t\partial_t\phi\right|\leq CC_1\varepsilon(s/t)^{1-3\delta}s^{-3/2+C(C_1\varepsilon)^{1/2}}$. Then, we have

$$\begin{aligned}
&\left\|L^{J_1}\underline{h}^{00}\partial^{I}L^{J_2}\partial_t\partial_t\phi\right\|_{L^2(\mathcal{H}^*_s)}\\
&\leq\left\|L^{J_1}\underline{h}^{00}_0\partial^{I}L^{J_2}\partial_t\partial_t\phi\right\|_{L^2(\mathcal{H}^*_s)}+\left\|L^{J_1}\underline{h}^{00}_1\partial^{I}L^{J_2}\partial_t\partial_t\phi\right\|_{L^2(\mathcal{H}^*_s)}\\
&\leq CC_1\varepsilon\left\|t^{-1}\partial^{I}L^{J_2}\partial_t\partial_t\phi\right\|_{L^2(\mathcal{H}^*_s)}+CC_1\varepsilon\left\|L^{J_1}\underline{h}^{00}_1\,(s/t)^{1-3\delta}s^{-3/2+C(C_1\varepsilon)^{1/2}}\right\|_{L^2(\mathcal{H}^*_s)}
\end{aligned}$$

thus

$$\begin{aligned}
\left\|L^{J_1}\underline{h}^{00}\partial^{I}L^{J_2}\partial_t\partial_t\phi\right\|_{L^2(\mathcal{H}^*_s)}&\leq CC_1s^{-1}\sum_{\substack{|I'|\leq|I|+1\\|J'|<|J|}}E_{M,c^2}(s,\partial^{I'}L^{J'}\phi)^{1/2}\\
&\quad+CC_1\varepsilon s^{-1/2+C(C_1\varepsilon)^{1/2}}\left\|s^{-1}(s/t)^{-1+\delta}L^{J_1}\underline{h}^{00}_1\right\|_{L^2(\mathcal{H}^*_s)}
\end{aligned}$$

and so

$$
\begin{aligned}
&\big\|L^{J_1}\underline{h}^{00}\partial^I L^{J_2}\partial_t\partial_t\phi\big\|_{L^2(\mathcal{H}^*_s)}\\
&\leq CC_1 s^{-1}\sum_{\substack{|I'|\leq|I|+1\\|J'|<|J|}} E_{M,c^2}(s,\partial^{I'}L^{J'}\phi)^{1/2}\\
&\quad + CC_1\varepsilon s^{-1/2+C(C_1\varepsilon)^{1/2}}\big\|s^{-1}(s/t)^{-1+\delta}L^{J_1}\underline{h}_1^{00}\big\|_{L^2(\mathcal{H}^*_s)}\\
&\quad + CC_1\varepsilon s^{-1/2+C(C_1\varepsilon)^{1/2}}\sum_{\substack{|J'|\leq|J|\\\alpha,\beta}}\int_2^s \tau^{-1}E^*_M(\tau,L^{J'}h_{\alpha\beta})^{1/2}d\tau\\
&\quad + C(C_1\varepsilon)^2 s^{-1/2+C(C_1\varepsilon)^{1/2}},
\end{aligned}
$$

where in the last inequality we applied Proposition 7.1.

• Case $1\leq J_1 = J$ then $|J_2| = 0$.

When $|J|\geq N-6$, we see that $|I|\leq 6\leq N-7$ provided by $N\geq 13$. In this case we apply (11.6) to $\partial^I L^{J_2}\partial_t\partial_t\phi$ and Proposition 7.1 on $L^{J_1}\underline{h}^{00}$:

$$
\begin{aligned}
&\big\|L^{J_1}\underline{h}^{00}\partial^I L^{J_2}\partial_t\partial_t\phi\big\|_{L^2(\mathcal{H}^*_s)} = \big\|L^{J}\underline{h}^{00}\partial^I\partial_t\partial_t\phi\big\|_{L^2(\mathcal{H}^*_s)}\\
&\leq CC_1\varepsilon\big\|t^{-1}\partial^I\partial_t\partial_t\phi\big\|_{L^2(\mathcal{H}^*_s)} + CC_1\varepsilon\Big\|(s/t)^{1-3\delta}s^{-3/2}L^J\underline{h}_1^{00}\Big\|_{L^2(\mathcal{H}^*_s)}.
\end{aligned}
$$

The first term is bounded by $CC_1\varepsilon s^{-1}\sum_{|I'|\leq|I|+1}E_{M,c^2}(\partial^{I'}\phi)^{1/2}$. For the second term, by applying Proposition 7.1, we have

$$
\begin{aligned}
&\Big\|(s/t)^{1-3\delta}s^{-3/2}L^J\underline{h}_1^{00}\Big\|_{L^2(\mathcal{H}^*_s)}\\
&\leq \Big\|(s/t)^{1-3\delta}s^{-3/2}s(s/t)^{1-\delta}\,s^{-1}(s/t)^{-1+\delta}L^J\underline{h}_1^{00}\Big\|_{L^2(\mathcal{H}^*_s)}\\
&\leq CC_1\varepsilon s^{-1/2}\sum_{\substack{|J_1'|\leq|J|\\\alpha,\beta}}E^*_M(s,L^{J_1'}h_{\alpha\beta})^{1/2}\\
&\quad + CC_1\varepsilon s^{-1/2}\sum_{\substack{|J_1'|\leq|J|\\\alpha,\beta}}\int_2^s\tau^{-1}E^*_M(\tau,L^{J_1'}h_{\alpha\beta})^{1/2}d\tau + C(C_1\varepsilon)^2 s^{-1/2}.
\end{aligned}
$$

When $|J|\leq N-7$, we apply (11.19) to $L^J\underline{h}^{00}$:

$$
\begin{aligned}
\big\|L^{J_1}\underline{h}^{00}\partial^I L^{J_2}\partial_t\partial_t\phi\big\|_{L^2(\mathcal{H}^*_s)} &\leq CC_1\varepsilon s^{-1+C(C_1\varepsilon)^{1/2}}\|(s/t)\partial^I\partial_t\partial_t\phi\|_{L^2(\mathcal{H}^*_s)}\\
&\leq CC_1\varepsilon s^{-1+C(C_1\varepsilon)^{1/2}}E_{M,c^2}(\partial^I\partial_t\phi)^{1/2}.
\end{aligned}
$$

We emphasize that such a term does not exist when $|J| = 0$ since the condition $1\leq|J_1|\leq|J|$ is then never satisfied.

Term 3. $\underline{h}^{00}\partial_\gamma\partial_{\gamma'}\partial^I L^{J'}$ with $|J'| < |J|$. This term is easier. We apply (11.16) to $\underline{h}^{00}$:

$$\begin{aligned}\left\|\underline{h}^{00}\partial_\gamma\partial_{\gamma'}\partial^I L^{J'}\phi\right\|_{L^2(\mathcal{H}^*_s)} &\leq CC_1\varepsilon s^{-1+C(C_1\varepsilon)^{1/2}}\left\|(s/t)\partial_\gamma\partial_{\gamma'}\partial^I L^{J'}\phi\right\|_{L^2(\mathcal{H}^*_s)}\\ &\leq CC_1\varepsilon s^{-1+C(C_1\varepsilon)^{1/2}}\sum_{\substack{|I'|\leq|I|+1\\|J'|<|J|}} E^*_M(s,\partial^{I'}L^{J'}\phi)^{1/2}.\end{aligned}$$

We now collect all the above estimates together and the desired result (12.4) is proved. Furthermore, when $|J|=0$, the condition $|J'|<|J|$ in the sum of the third, the fourth and fifth term in the right-hand side of (12.4) indicate that these three terms disappear. Furthermore, when $|J|=0$, the term $L^{J_1}\underline{h}^{00}\partial^I L^{J_2}\partial_t\partial_t\phi$ and $\underline{h}^{00}\partial_\gamma\partial_{\gamma'}\partial^I L^{J'}$ do not exist (since they demand $|J_1|\geq 1$ and $|J'|<|J|$). So, the only existent terms are $\partial^{I_1}\underline{h}^{00}\partial^{I_2}\partial_t\partial_t\phi$, the null terms and the commutative terms with additional t^{-1} decay. They can be bounded by $C(C_1\varepsilon)^2 s^{-1+2\delta}$ and this concludes the derivation of (12.5). □

12.2 Main estimates in this section

Proposition 12.1. *Let the bootstrap assumptions* (5.1) *and* (5.2) *hold with C_1/C_0 sufficiently large, then there exists a positive constant ε_3 sufficiently small so that for all $\varepsilon\leq\varepsilon_3$ and for $N-3\leq|I|+|J|\leq N$*

$$E^*_M(s,\partial^I L^J h_{\alpha\beta})^{1/2}\leq\frac{1}{2}C_1\varepsilon s^{C(C_1\varepsilon)^{1/2}},\tag{12.7}$$

$$E_{M,c^2}(s,\partial^I L^J\phi)^{1/2}\leq\frac{1}{2}C_1\varepsilon s^{1/2+C(C_1\varepsilon)^{1/2}}.\tag{12.8}$$

The proof will be split into two parts. First, we will derive (12.7) and (12.8) in the case $|J|=0$. In a second part, we will propose an induction argument for the case $|J|\neq 0$.

Proof of Proposition 12.1 in the case $|J|=0$. In this case, the following estimates are direct by Lemma 10.1, (12.1), (12.2) and (12.4):

$$\|\partial^I F_{\alpha\beta}\|_{L^2(\mathcal{H}^*_s)}\leq CC_1\varepsilon s^{-1}\sum_{\substack{|I'|\leq|I|\\\alpha',\beta'}} E^*_M\big(s,\partial^{I'}h_{\alpha'\beta'}\big)^{1/2}+C(C_1\varepsilon)^2 s^{-3/2+2\delta}$$

and

$$\begin{aligned}&\|\partial^I(\partial_\alpha\phi\partial_\beta\phi)\|_{L^2(\mathcal{H}_s^*)}+\|\partial^I(\phi^2)\|_{L^2(\mathcal{H}_s^*)}\\&\leq C(C_1\varepsilon)s^{-3/2}\sum_{|I'|\leq|I|}E_{M,c^2}(s,\partial^{I'}\phi)^{1/2}\\&\leq C(C_1\varepsilon)^2s^{-3/2+\delta}+C(C_1\varepsilon)s^{-3/2}\sum_{N-3\leq|I'|\leq|I|}E_{M,c^2}(s,\partial^{I'}\phi)^{1/2},\end{aligned}$$

while

$$\begin{aligned}&\|[\partial^I,h^{\mu\nu}\partial_\mu\partial_\nu]h_{\alpha\beta}\|_{L^2(\mathcal{H}_s^*)}\\&\leq C(C_1\varepsilon)^2s^{-3/2+3\delta}+CC_1\varepsilon s^{-3/2}\sum_{N-3\leq|I'|\leq|I|}E_{M,c^2}(s,\partial^{I'}L^J\phi)^{1/2}\end{aligned}$$

and

$$\left\|[\partial^I,h^{\mu\nu}\partial_\mu\partial_\nu]\phi\right\|_{L^2_f(\mathcal{H}_s)}\leq C(C_1\varepsilon)^2s^{-1+3\delta}.$$

And by Lemma 7.3, we obtain $M_{\alpha\beta}[\partial^IL^Jh](s)\leq C(C_1\varepsilon)^2s^{-3/2+2\delta}$ and

$$M[\partial^IL^J\phi](s)\leq C(C_1\varepsilon)^2s^{-1+2\delta}.$$

We conclude that in view of (3.10) and (3.2) (by observe that (3.1) is guaranteed by Lemma 7.2):

$$E_{M,c^2}(s,\partial^I\phi)^{1/2}\leq CC_0\,\varepsilon+C(C_1\varepsilon)^2s^{2\delta},\tag{12.9}$$

$$\begin{aligned}E_M^*(s,\partial^Ih_{\alpha\beta})^{1/2}&\leq CC_0\,\varepsilon+C(C_1\varepsilon)^2\\&\quad+CC_1\varepsilon\sum_{\substack{|I'|\leq|I|\\\alpha',\beta'}}\int_2^s\tau^{-1}E_M^*\big(\tau,\partial^{I'}h_{\alpha'\beta'}\big)^{1/2}d\tau\\&\quad+CC_1\varepsilon\sum_{N-3\leq|I'|\leq|I|}\int_2^s\tau^{-3/2}E_{M,c^2}(\tau,\partial^{I'}\phi)^{1/2}d\tau.\end{aligned}\tag{12.10}$$

Substituting (12.9) into (12.10), we obtain

$$\begin{aligned}&E_M^*(s,\partial^Ih_{\alpha\beta})^{1/2}\\&\leq CC_0\,\varepsilon+C(C_1\varepsilon)^2+CC_1\varepsilon\sum_{\substack{|I'|\leq|I|\\\alpha',\beta'}}\int_2^s\tau^{-1}E_M^*\big(\tau,\partial^{I'}h_{\alpha'\beta'}\big)^{1/2}d\tau.\end{aligned}\tag{12.11}$$

Now, in view of (12.11), we introduce the notation $Y(s):=\sum_{\substack{|I|\leq N\\\alpha,\beta}}E_M^*(s,\partial^Ih_{\alpha\beta})^{1/2}$. With this notation, the estimate (12.11) transforms into

$$Y(s)\leq CC_0\,\varepsilon+C(C_1\varepsilon)^2+CC_1\varepsilon\int_2^s\tau^{-1}Y(\tau)d\tau.\tag{12.12}$$

Then Gronwall's inequality leads us to

$$\sum_{\substack{|I|\leq N\\ \alpha,\beta}} E_M(s,\partial^{I'}h_{\alpha\beta})^{1/2} = Y(s) \leq C(C_0\,\varepsilon + (C_1\varepsilon)^2)s^{CC_1\varepsilon}. \tag{12.13}$$

In (12.9) and (12.13), we take $\varepsilon_{20} = \frac{C_1-2CC_0}{2C_1^2}$ and for all $0\leq\varepsilon\leq\varepsilon_{20}$, we obtain

$$E_M(s,\partial^I h_{\alpha\beta})^{1/2} \leq \frac{1}{2}C_1\varepsilon s^{CC_1\varepsilon}$$

and

$$E_{M,c^2}(s,\partial^I h_{\alpha\beta})^{1/2} \leq \frac{1}{2}C_1\varepsilon s^{CC_1\varepsilon}.$$

This yields the desired result for $|J|=0$. □

Proof of Proposition 12.1, Case $1\leq|J|\leq N$. We proceed by induction on $|J|$ and assume that for $|I|+|J'|\leq N-1$ and $|J'|\leq m-1<N$

$$\begin{aligned}
&E_M(s,\partial^I L^{J'}h_{\alpha\beta})^{1/2} \leq C(C_0\,\varepsilon + (C_1\varepsilon)^2)s^{C(C_1\varepsilon)^{1/2}},\\
&E_{M,c^2}(s,\partial^I L^{J'}\phi)^{1/2} \leq C(C_0\,\varepsilon + (C_1\varepsilon))^2 s^{1/2+C(C_1\varepsilon)^{1/2}}.
\end{aligned} \tag{12.14}$$

We will prove that it is again valid for $|J|=m\leq N$ by using Propositions 3.1 and 3.2. From the induction assumption, we have

$$\begin{aligned}
\|\partial^I L^J F_{\alpha\beta}\|_{L^2(\mathcal{H}_s^*)} &\leq CC_1\varepsilon s^{-1}\sum_{\substack{|I'|\leq|I|\\ \alpha,\beta}} E_M^*(s,\partial^{I'}L^J h_{\alpha\beta})^{1/2}\\
&\quad + CC_1\varepsilon\left(C_0\,\varepsilon + (C_1\varepsilon)^2\right)s^{-1+C(C_1\varepsilon)^{1/2}}
\end{aligned}$$

thanks to (10.1),

$$\begin{aligned}
&\left\|\partial^I L^J\left(\partial_\alpha\phi\partial_\beta\phi\right)\right\|_{L^2(\mathcal{H}_s^*)} + \left\|\partial^I L^J\left(\phi^2\right)\right\|_{L^2(\mathcal{H}_s^*)}\\
&\leq CC_1\varepsilon s^{-3/2}\sum_{|I'|\leq|I|} E_{M,c^2}(s,\partial^{I'}L^J\phi)^{1/2} + CC_1\varepsilon\left(C_0\,\varepsilon + (C_1\varepsilon)^2\right)s^{-1+C(C_1\varepsilon)^{1/2}}
\end{aligned}$$

thanks to (12.1), and finally in view of (12.2).

$$\begin{aligned}
&\left\|[\partial^I L^J, h^{\mu\nu}\partial_\mu\partial_\nu]h_{\alpha\beta}\right\|_{L^2(\mathcal{H}_s^*)}\\
&\leq CC_1\varepsilon s^{-1}\sum_{\substack{|J'|=|J|\\ |I'|\leq|I|}} E_M^*(s,\partial^I L^{J'}h_{\alpha\beta})^{1/2} + CC_1\varepsilon\left(C_0\,\varepsilon + (C_1\varepsilon)^2\right)s^{-1+C(C_1\varepsilon)^{1/2}}.
\end{aligned}$$

On the other hand, in view of (12.4), we have

$$\begin{aligned}
&\big\|[\partial^I L^J, h^{\mu\nu}\partial_\mu\partial_\nu]\phi\big\|_{L^2(\mathcal{H}_s^*)} \\
&\leq CC_1\varepsilon s^{-1/2} \sum_{\substack{|J'|=|J| \\ \alpha,\beta}} E_M^*(s, L^{J'} h_{\alpha\beta})^{1/2} \\
&\quad + CC_1\varepsilon s^{-1/2} \sum_{\substack{|J'|=|J| \\ \alpha,\beta}} \int_2^s \tau^{-1} E_M^*(\tau, J^{J'} h_{\alpha\beta})^{1/2} \\
&\quad + CC_1\varepsilon \left(C_0 + (C_1\varepsilon)^2\right) s^{-1/2+C(C_1\varepsilon)^{1/2}} \\
&\quad + CC_1\varepsilon \left(C_0\,\varepsilon + (C_1\varepsilon)^2\right) s^{-1/2+C(C_1\varepsilon)^{1/2}} \int_2^s \tau^{-1+C(C_1\varepsilon)^{1/2}} d\tau \\
&\quad + C(C_1\varepsilon)^2 s^{-1/2+C(C_1\varepsilon)^{1/2}},
\end{aligned}$$

thus

$$\begin{aligned}
&\big\|[\partial^I L^J, h^{\mu\nu}\partial_\mu\partial_\nu]\phi\big\|_{L^2(\mathcal{H}_s^*)} \\
&\leq CC_1\varepsilon s^{-1/2} \sum_{\substack{|J'|=|J| \\ \alpha,\beta}} E_M^*(s, L^{J'} h_{\alpha\beta})^{1/2} \\
&\quad + CC_1\varepsilon s^{-1/2} \sum_{\substack{|J'|=|J| \\ \alpha,\beta}} \int_2^s \tau^{-1} E_M^*(\tau, J^{J'} h_{\alpha\beta})^{1/2} \\
&\quad + C(C_1\varepsilon)^2 s^{-1/2+C(C_1\varepsilon)^{1/2}}.
\end{aligned}$$

Also, in view of (7.6) we have $M_{\alpha\beta}[\partial^I L^J h] \leq C(C_1\varepsilon)^2 s^{-3/2+2\delta}$ for $|I|+|J| \leq N$.

With

$$W_m(s) := \sum_{\substack{|J|=m,\alpha,\beta \\ |I|+|J|\leq N}} E_M(s, \partial^I L^J h_{\alpha\beta})^{1/2}$$

and

$$K_m(s) := s^{-1/2} \sum_{\substack{|J|=m \\ |I|+|J|\leq N}} E_{M,c^2}(s, \partial^I L^J \phi)^{1/2},$$

the energy estimates (3.2) and (3.10) lead us to a system of integral inequalities:

$$\begin{aligned}
W_m(s) &\leq C\left(C_0\,\varepsilon + (C_1\varepsilon)^2\right) s^{C(C_1\varepsilon)^{1/2}} + CC_1\varepsilon \int_2^s \tau^{-1}\left(W_m(\tau) + K_m(\tau)\right) d\tau \\
K_m(s) &\leq C\left(C_0\,\varepsilon + (C_1\varepsilon)^2\right) s^{C(C_1\varepsilon)^{1/2}} + CC_1\varepsilon s^{-1/2} \int_2^s \tau^{-1/2} W_m(\tau)\, d\tau \\
&\quad + CC_1\varepsilon s^{-1/2} \int_2^s \tau^{-1/2} \int_2^\tau \eta^{-1} W_m(\eta) d\eta d\tau.
\end{aligned} \tag{12.15}$$

Lemma 12.4 stated and proven below will guarantee that (12.15) leads us

$$W_m(s) + K_m(s) \leq C\left(C_0\,\varepsilon + (C_1\varepsilon)^2\right) s^{C(C_1\varepsilon)^{1/2}}.$$

This leads us to the desired $|J| = m$ case. Then, by induction, (12.7) is valid for all $|J| = m \leq N$. We see that we can choose $\varepsilon_3 := \frac{C_1 - 2CC_0}{2CC_1^2}$ with $C_1 > 2CC_0$, then

$$W_m(s) + K_m(s) \leq \frac{1}{2} C_1 \varepsilon s^{C(C_1\varepsilon)^{1/2}}$$

for $0 \leq \varepsilon \leq \varepsilon_3$. This concludes the proof of Proposition 12.1. □

Lemma 12.4. *Let W and K be positive, locally integrable functions defined in $[0,T]$, and suppose that*

$$\begin{aligned} W(s) &\leq C\left(C_0\,\varepsilon + (C_1\varepsilon)^2\right) s^{C(C_1\varepsilon)^{1/2}} + CC_1\varepsilon \int_2^s \tau^{-1}\left(W(\tau) + K(\tau)\right) d\tau, \\ K(s) &\leq C\left(C_0\,\varepsilon + (C_1\varepsilon)^2\right) s^{C(C_1\varepsilon)^{1/2}} + CC_1\varepsilon s^{-1/2} \int_2^s \tau^{-1/2} W(\tau)\, d\tau \\ &\quad + CC_1\varepsilon s^{-1/2} \int_2^s \tau^{-1/2} \int_2^\tau \eta^{-1} W(\eta) d\eta d\tau \end{aligned} \tag{12.16}$$

hold for some constant $C > 0$ and a sufficiently small constant $C_1\varepsilon > 0$. Then, one has

$$W(s) + K(s) \leq C\left(C_1\varepsilon + (C_1\varepsilon)^2\right) s^{C(C_1\varepsilon)^{1/2}}, \quad s \in [0,T].$$

Proof. We define

$$W^*(s) := \sup_{\tau\in[0,s]} \left\{\tau^{-C(C_1\varepsilon)^{1/2}} W(\tau)\right\}$$

as well as

$$K^*(s) := \sup_{s\in[0,s]} \left\{\tau^{-C(C_1\varepsilon)^{1/2}} K(\tau)\right\}.$$

With this notation, (12.16) yields us to (after taking the supremum over s)

$$\begin{aligned} W^*(s) \leq& C\left(C_0\,\varepsilon + (C_1\varepsilon)^2\right) \\ &+ CC_1\varepsilon s^{-C(C_1\varepsilon)^{1/2}} \left(W^*(s) + K^*(s)\right) \int_2^s \tau^{-1+C(C_1\varepsilon)^{1/2}} d\tau, \end{aligned}$$

which leads us to

$$W^*(s) \leq C\left(C_0\,\varepsilon + (C_1\varepsilon)^2\right) + C(C_1\varepsilon)^{1/2}\left(W^*(s) + K^*(s)\right).$$

A similar argument can be applied to estimate K and we also find

$$K^*(s) \leq C\left(C_0\,\varepsilon + (C_1\varepsilon)^2\right) + CC_1\varepsilon W^*(s) + C(C_1\varepsilon)^{1/2} W^*(s). \tag{12.17}$$

By taking the sum of the above two estimates and when $(C_1\varepsilon)$ is sufficiently small, there exists a constant $\varepsilon_4 > 0$, such that if $\varepsilon \leq C_1^{-1}\varepsilon_4$,

$$W^*(s) + K^*(s) \leq C\left(C_0\,\varepsilon + (C_1\varepsilon)^2\right) + C(C_1\varepsilon)^{1/2}\left(W^*(s) + K^*(s)\right). \tag{12.18}$$

Since $C(C_1\varepsilon)^{1/2} \leq 1/2$ (for $C_1\varepsilon$ sufficiently small) we have

$$W^*(s) + K^*(s) \leq C\left(C_0\,\varepsilon + (C_1\varepsilon)^2\right),$$

which leads us to the desired result. □

12.3 Applications to the derivation of refined decay estimates

With the refined energy at higher-order, we can establish some additional refined decay estimates. This subsection is totally parallel to Section 10.3. First, by the global Sobolev inequality, for $|I| + |J| \leq N - 2$:

$$|\partial^I L^J \partial_\gamma h_{\alpha\beta}| + |\partial_\gamma \partial^I L^J h_{\alpha\beta}| \leq CC_1\varepsilon t^{-1/2} s^{-1+C(C_1\varepsilon)^{1/2}}, \tag{12.19}$$

$$|\partial^I L^J \underline{\partial}_a h_{\alpha\beta}| + |\underline{\partial}_a \partial^I L^J h_{\alpha\beta}| \leq CC_1\varepsilon t^{-3/2} s^{C(C_1\varepsilon)^{1/2}}. \tag{12.20}$$

Based on this improved sup-norm estimate, the following estimates are direct by integration along the rays $\{(t, \lambda x)|1 \leq \lambda \leq t/|x|\}$:

$$|\partial^I L^J h_{\alpha\beta}| \leq CC_1\varepsilon\left(t^{-1} + (s/t)t^{-1/2} s^{C(C_1\varepsilon)^{1/2}}\right). \tag{12.21}$$

From the above estimates and Lemma 4.7, we have

$$\left|\partial^I L^J \partial_\alpha \underline{h}^{00}\right| + \left|\partial^I L^J \partial_\alpha \underline{h}^{00}\right| \leq CC_1\varepsilon t^{-3/2} s^{C(C_1\varepsilon)^{1/2}} \tag{12.22}$$

and also by integration along the rays $\{(t, \lambda x)|1 \leq \lambda \leq t/|x|\}$:

$$\left|\partial^I L^J \underline{h}^{00}\right| \leq CC_1\varepsilon\left(t^{-1} + (s/t)^2 t^{-1/2} s^{C(C_1\varepsilon)^{1/2}}\right). \tag{12.23}$$

Two more delicate applications of this higher-order, improved energy estimate are discussed in the following. They are also parallel to Lemmas 10.3 and 10.4.

Lemma 12.5. *For $|I| + |J| \leq N - 2$, one has*

$$\left|\partial^I L^J F_{\alpha\beta}\right| \leq C(C_1\varepsilon)^2 t^{-1} s^{-2+C(C_1\varepsilon)^{1/2}}. \tag{12.24}$$

Proof. We focus on $F_{\alpha\beta}$. Recall that $F_{\alpha\beta} = Q_{\alpha\beta} + P_{\alpha\beta}$. We see that (omit cubic and higher-order terms, which have good decay), the quadratic part of $F_{\alpha\beta}$ are linear combinations of $\partial_\gamma h_{\alpha\beta}\partial_{\gamma'} h_{\alpha'\beta'}$. Then, we apply (12.19) and see that, for $|I| + |J| \leq N - 2$, we find $\partial^I L^J\left(\partial_\gamma h_{\alpha\beta}\partial_{\gamma'} h_{\alpha'\beta'}\right) \leq C(C_1\varepsilon)^2 t^{-1} s^{-2+C(C_1\varepsilon)^{1/2}}$. □

A second refined estimate parallel to Lemma 10.4 can now be derived. The proof is essentially the same as that of Lemma 10.4. The only difference is that we apply the sup-norm estimates presented in Lemma 12.5 for $|I|+|J| \leq N-2$.

Lemma 12.6. *For $|I|+|J| \leq N-3$, one has*

$$\left|\partial_t\partial_t\partial^I L^J h_{\alpha\beta}\right| \leq CC_1\varepsilon t^{1/2}s^{-3+(CC_1\varepsilon)^{1/2}}. \tag{12.25}$$

By a similar argument as done below (7.23), we have

$$|\partial_\alpha\partial_\beta\partial^I L^J h_{\alpha\beta}| + |\partial^I L^J\partial_\alpha\partial_\beta h_{\alpha\beta}| \leq CC_1\varepsilon t^{1/2}s^{-3+(CC_1\varepsilon)^{1/2}}. \tag{12.26}$$

Apart from the above refined decay on $h_{\alpha\beta}$, we also have the following refined decay for ϕ, deduced from (12.8). For $|I|+|J| \leq N-2$, we have

$$\begin{aligned} \left|\partial^I L^J\partial_\alpha\phi\right| + \left|\partial_\alpha\partial^I L^J\phi\right| &\leq CC_1\varepsilon t^{-1/2}s^{-1/2+C(C_1\varepsilon)^{1/2}}, \\ \left|\partial^I L^J\underline{\partial}_a\phi\right| + \left|\underline{\partial}_a\partial^I L^J\phi\right| + \left|\partial^I L^J\phi\right| &\leq CC_1\varepsilon t^{-3/2}s^{1/2+C(C_1\varepsilon)^{1/2}}, \end{aligned} \tag{12.27}$$

while, for $|I|+|J| \leq N-3$, we apply (4.17) and get

$$\left|\partial^I L^J\underline{\partial}_a\phi\right| + \left|\underline{\partial}_a\partial^I L^J\phi\right| \leq CC_1\varepsilon t^{-5/2}s^{1/2+C(C_1\varepsilon)^{1/2}}. \tag{12.28}$$

Finally, for $|I|+|J| \leq N-4$, we have

$$\left|\partial^I L^J\partial_\beta\underline{\partial}_a\phi\right| + \left|\underline{\partial}_a\partial_\beta\partial^I L^J\phi\right| \leq CC_1\varepsilon t^{-5/2}s^{1/2+C(C_1\varepsilon)^{1/2}}, \tag{12.29}$$

$$\left|\partial_\alpha\partial_\beta\partial^I L^J\phi\right| + \left|\partial^I L^J\partial_\alpha\partial_\beta\phi\right| \leq CC_1\varepsilon t^{-3/2}s^{1/2+C(C_1\varepsilon)^{1/2}}. \tag{12.30}$$

Chapter 13

High-Order Refined Sup-Norm Estimates

13.1 Preliminary

We begin with our refined estimates for $\partial^I L^J (h^{\mu\nu}\partial_\mu\partial_\nu h_{\alpha\beta})$, QS_ϕ and $[\partial^I L^J, h^{\mu\nu}\partial_\mu\partial_\nu]\phi$ for $|I|+|J| \le N-4$.

Lemma 13.1. *For all $|I|+|J| \le N-4$, the following estimate holds:*

$$\left|L^J (h^{\mu\nu}\partial_\mu\partial_\nu h_{\alpha\beta})\right| \le C(C_1\varepsilon)^2 t^{-2+C(C_1\varepsilon)^{1/2}}(t-r)^{-1+C(C_1\varepsilon)^{1/2}}. \tag{13.1}$$

Proof. The proof is parallel to that of Lemma 11.1. The only difference is that there we only have refined decay estimates on $\partial^I L^J \partial_t\partial_t h_{\alpha\beta}$ and $L^J \underline{h}^{00}$ for $|I|+|J| \le 7$ but here we have, in view of (12.25) and (12.26), the parallel estimate for $|I|+|J| \le N-3$. □

Lemma 13.2. *For $|I|+|J| \le N-4$, the following estimate holds:*

$$\begin{aligned}
&\left|[\partial^I L^J, h^{\mu\nu}\partial_\nu\partial_\nu]\phi\right| \\
&\le C(C_1\varepsilon)^2 (s/t)^3 s^{-3+2\delta} + CC_1\varepsilon (s/t)^{3/2} s^{-3/2+\delta} \sum_{\substack{|I_2|\le|I|-1\\|J_2|\le|J|}} \left|\partial_t\partial_t\partial^{I_2}L^{J_2}\phi\right| \\
&\quad + CC_1\varepsilon t^{-1} s^{C(C_1\varepsilon)^{1/2}} \sum_{|J'|<|J|} \left|\partial_t\partial_t\partial^I L^{J'}\phi\right| \\
&\quad + CC_1\varepsilon (s/t)^{1-3\delta} s^{-3/2+C(C_1\varepsilon)^{1/2}} \sum_{\substack{|J'|<|J|,\\ \alpha,\beta}} \left|L^{J'}h_{\alpha\beta}\right| \\
&\quad + CC_1\varepsilon (s/t)^{1-3\delta} s^{-3/2} \sum_{\alpha,\beta} \left|L^J h_{\alpha\beta}\right|
\end{aligned} \tag{13.2}$$

and, when $|J| = 0$,

$$
\begin{aligned}
&\big|[\partial^I, h^{\mu\nu}\partial_\nu\partial_\nu]\phi\big| \\
&\leq C(C_1\varepsilon)^2(s/t)^3 s^{-3+2\delta} + CC_1\varepsilon(s/t)^{3/2}s^{-3/2+\delta} \sum_{|I_2|\leq|I|-1} \big|\partial_t\partial_t\partial^{I_2}\phi\big| .
\end{aligned}
\tag{13.3}
$$

Proof. The proof relies on the decomposition presented in (4.16) combined with the refined decay estimates on ∂h, ϕ and $\partial\phi$ presented in Section 12.3. We see that the null terms and the terms of commutators listed in (4.16) are bounded by trivial application of the refined decay estimates presented in Section 12.3. We only write the estimate on the null term $\partial^{I_1}L^{J_1}\underline{h}^{a0}\partial^{I_2}L^{J_2}\underline{\partial}_a\partial_t\phi$ (and omit the treatment of the other terms). We see that $\underline{h}^{a0}$ is a linear combination of $h_{\alpha\beta}$ with smooth and homogeneous coefficients plus higher-order correction terms:

Case 1. When $|I_1| \geq 1$, we apply the basic sup-norm estimates (5.12a) and (4.18):

$$
\begin{aligned}
\big|\partial^{I_1}L^{J_1}\underline{h}^{a0}\partial^{I_2}L^{J_2}\underline{\partial}_a\partial_t\phi\big| &\leq CC_1\varepsilon t^{-1/2}s^{-1+\delta}\, CC_1\varepsilon t^{-3/2}s^{1/2+\delta} \\
&\leq C(C_1\varepsilon)^2(s/t)^2 s^{-5/2+2\delta}.
\end{aligned}
$$

Case 2. When $|I_1| = 0$, we apply (5.22) and (4.18):

$$
\begin{aligned}
&\big|\partial^{I_1}L^{J_1}\underline{h}^{a0}\partial^{I_2}L^{J_2}\underline{\partial}_a\partial_t\phi\big| = \big|L^{J_1}\underline{h}^{a0}\partial^{I}L^{J_2}\underline{\partial}_a\partial_t\phi\big| \\
&\leq CC_1\varepsilon\left((s/t)t^{-1/2}s^{\delta} + t^{-1}\right)CC_1\varepsilon t^{-5/2}s^{1/2+\delta} \leq C(C_1\varepsilon)^2(s/t)^4 s^{-5/2+2\delta}.
\end{aligned}
$$

We then focus on the estimates of the last three terms.

• We treat first the term $\partial^{I_1}L^{J_1}\underline{h}^{00}\partial^{I_2}L^{J_2}\partial_t\partial_t\phi$ with $|I_1| \geq 1$. We apply the sharp estimate to $\partial^{I_1}L^{J_1}\underline{h}^{00}$ provided by (7.1):

$$
\big|\partial^{I_1}L^{J_1}\underline{h}^{00}\partial^{I_2}L^{J_2}\partial_t\partial_t\phi\big| \leq CC_1\varepsilon(s/t)^{3/2}s^{-3/2+\delta} \sum_{\substack{|I_2|\leq|I|\\|J_2|\leq|J|}} \big|\partial^{I_2}L^{J_2}\partial_t\partial_t\phi\big| .
$$

By the commutator estimate (3.54), we have

$$
\big|\partial^{I_2}L^{J_2}\partial_t\partial_t\phi\big| \leq C \sum_{|J_2'|\leq|J_2|} \big|\partial_\gamma\partial_{\gamma'}\partial^I L^{J_2'}\phi\big| .
$$

Then we rely on the decomposition (7.23) and a similar argument and obtain

$$
\big|\partial_\gamma\partial_{\gamma'}\partial^I L^{J_2'}\phi\big| \leq \big|\partial_t\partial_t\partial^I L^{J_2'}\phi\big| + CC_1\varepsilon t^{-5/2}s^{1/2+\delta},
$$

so that

$$
\begin{aligned}
&\big|\partial^{I_1}L^{J_1}\underline{h}^{00}\partial^{I_2}L^{J_2}\partial_t\partial_t\phi\big| \\
&\leq CC_1\varepsilon(s/t)^{3/2}s^{-3/2+\delta} \sum_{\substack{|I_2|\leq|I|-1\\|J_2|\leq|J|}} \big|\partial_t\partial_t\partial^{I_2}L^{J_2}\phi\big| + C(C_1\varepsilon)^2(s/t)^4 s^{-7/2+2\delta}.
\end{aligned}
$$

• The term $L^{J_1'}\underline{h}^{00}\partial^I L^{J_2'}\phi$ is bounded as follows. We see that $|J_2'|<|J|$ and we will discuss the following cases:
Case 1. When $1\le |J_1'|\le N-7$, we apply (11.19):

$$\left|L^{J_1'}\underline{h}^{00}\partial^I L^{J_2'}\partial_t\partial_t\phi\right|\le CC_1\varepsilon t^{-1}s^{C(C_1\varepsilon)^{1/2}}\,CC_1\varepsilon\sum_{|J'|<|J|}\left|\partial^I L^{J'}\partial_t\partial_t\phi\right|.$$

Applying the same estimate for $\left|\partial^I L^{J'}\partial_t\partial_t\phi\right|$ as above, we conclude that

$$\begin{aligned}&\left|L^{J_1'}\underline{h}^{00}\partial^I L^{J_2'}\partial_t\partial_t\phi\right|\\&\le CC_1\varepsilon t^{-1}s^{C(C_1\varepsilon)^{1/2}}\sum_{|J'|<|J|}\left|\partial_t\partial_t\partial^I L^{J'}\phi\right|+C(C_1\varepsilon)^2(s/t)^{7/2}s^{-3+C(C_1\varepsilon)^{1/2}}.\end{aligned}$$

Case 2. When $N-6\le |J_1'|\le |J|-1$, we have $|I|+|J_2'|\le 2\le N-8$, then we apply the last inequality of (11.9) to $\partial^I L^{J_2'}\partial_t\partial_t\phi$:

$$\left|L^{J_1'}\underline{h}^{00}\partial^I L^{J_2'}\partial_t\partial_t\phi\right|\le CC_1\varepsilon(s/t)^{1-3\delta}s^{-3/2+C(C_1\varepsilon)^{1/2}}\sum_{\substack{|J'|<|J|,\\ \alpha,\beta}}\left|L^{J'}h_{\alpha\beta}\right|.$$

Case 3. When $N-6\le |J_1'|$ and $J_1'=J$, we have $|I|\le 2\le N-8$ and $|J_2'|=0$. We apply (11.6):

$$\left|L^{J_1'}\underline{h}^{00}\partial^I L^{J_2'}\partial_t\partial_t\phi\right|=\left|L^J\underline{h}^{00}\partial^I\partial_t\partial_t\phi\right|\le CC_1\varepsilon(s/t)^{1-3\delta}s^{-3/2}\sum_{\alpha,\beta}\left|L^J h_{\alpha\beta}\right|.$$

The term $\underline{h}^{00}\partial_\gamma\partial_{\gamma'}\partial^I L^{J'}\phi$ is bounded by

$$CC_1\varepsilon t^{-1}s^{C(C_1\varepsilon)^{1/2}}\sum_{|J'|<|J|}\left|\partial_t\partial_t\partial^I L^{J'}\phi\right|+C(C_1\varepsilon)^2(s/t)^{7/2}s^{-3+C(C_1\varepsilon)^{1/2}}.$$

We omit the details of the proof which are essentially the same as in *Case 1* for $\partial^{I_1}L^{J_1}\underline{h}^{00}\partial^{I_2}L^{J_2}\phi$. Therefore, we have established (13.2).

For (13.3), when $|J|=0$, the third and fourth terms in the right-hand side of (13.2) disappear. The last term also disappear since, if we follow the proof of (13.2), we see that when $|J|=0$, and the *case 3* of $L^{J_1'}\underline{h}^{00}\partial^I L^{J_2'}\phi$ does not exist ($N-6\le J_1'$ and $J_1=J$ is contradictory). This is the only place that the last term in the right-hand side of (13.2) appears. Therefore, we have established (13.3). □

13.2 Main estimate in this section

Proposition 13.1. *There exist constants $C_1,\varepsilon_4>0$ such that if the bootstrap condition* (5.1)-(5.2) *holds with $C_1>C_0$ sufficiently large, then there* *exists a constant* $\varepsilon_4 > 0$ *such that for any* $\varepsilon \in (0, \varepsilon_4)$ *and* $N-6 \leq |I|+|J| \leq N-4$:

$$|L^J h_{\alpha\beta}| \leq CC_1\varepsilon t^{-1} s^{C(C_1\varepsilon)^{1/2}}, \tag{13.4}$$

$$(s/t)^{3\delta-2}|\partial^I L^J \phi| + (s/t)^{3\delta-3}|\partial^I L^J \underline{\partial}_\perp \phi| \leq CC_1\varepsilon t^{-3/2} s^{C(C_1\varepsilon)^{1/2}}. \tag{13.5}$$

The proof is divided into two parts and we analyze first the case $|J| = 0$.

Proof of Proposition 13.1 in the case $|J| = 0$**.** We see that (13.4) is already guaranteed by (11.16). To establish (13.5), we rely on Proposition 3.5 and follow the notation therein. The terms R_i are already bounded by Lemma 11.3, while the commutator term $[\partial^I, h^{\mu\nu}\partial_\mu\partial_\nu]\phi$ is bounded in view of (13.3). Hence, we have (always with $s = \sqrt{t^2 - r^2}$)

$$\begin{aligned}
F(t,x) \leq\, & CC_1\varepsilon(s/t)^{3/2} \int_{s_0}^{s} \tau^{-3/2+3\delta} d\tau + C(C_1\varepsilon)^2 (s/t)^3 \int_{s_0}^{s} \tau^{-3+2\delta}\, \tau^{3/2} d\tau \\
& + CC_1\varepsilon(s/t)^{3/2} \sum_{|I'|\leq|I|-1} \int_{s_0}^{s} \lambda^\delta \left|\partial^{I'}\partial_t\partial_t\phi\right| (\lambda t/s, \lambda x/s) d\lambda,
\end{aligned}$$

so

$$\begin{aligned}
& F(t,x) \\
& \leq CC_1\varepsilon(s/t)^{3/2} s_0^{-1/2+3\delta} + C(C_1\varepsilon)^2 (s/t)^3 \\
& \quad + CC_1\varepsilon(s/t)^{3/2} \sum_{|I'|\leq|I|-1} \int_{s_0}^{s} \lambda^\delta \left|\partial^{I'}\partial_t\partial_t\phi\right| (\lambda t/s, \lambda x/s) d\lambda \\
& \leq CC_1\varepsilon(s/t)^{2-3\delta} + CC_1\varepsilon(s/t)^{3/2} \sum_{|I'|\leq|I|-1} \int_{s_0}^{s} \lambda^\delta \left|\partial^{I'}\partial_t\partial_t\phi\right| (\lambda t/s, \lambda x/s) d\lambda,
\end{aligned}$$

where we recall that $s_0 \simeq \frac{t}{s}$.

Setting

$$X_n(\tau) := \sum_{|I|\leq n} \sup_{\mathcal{K}_{[2,\tau]}} \Big((s/t)^{3\delta-2} s^{3/2} \left|\partial^I\phi\right| + (s/t)^{3\delta-3} s^{3/2} \left|\underline{\partial}_\perp \partial^I \phi\right| \Big)(t,x),$$

we claim that

$$\left|(s/t)^{3\delta-1}\partial^{I'}\partial_t\partial_t\phi\right| (t,x) \leq Cs^{-3/2} X_n(s) + Ct^{-1}\epsilon(s/t)^{3\delta-1/2} s^{-1/2+\delta}, \tag{13.6}$$

which will be explained at the end of this proof. Replacing t by $\lambda t/s$ and integrating in λ, we then obtain

$$
\begin{aligned}
&F(t,x)\\
&\le C(C_1\varepsilon)(s/t)^{2-3\delta}\\
&\quad + CC_1\varepsilon(s/t)^{5/2-3\delta}\int_{s_0}^{s}\Big(\lambda^{-3/2+\delta}X_n(\lambda)+\epsilon(s/t)^{3\delta+1/2}\lambda^{-3/2+2\delta}\Big)\,d\lambda\\
&\le C(C_1\varepsilon)(s/t)^{2-3\delta}\\
&\quad + CC_1\varepsilon(s/t)^{5/2-3\delta}\Big(X_n(s)\int_{s_0}^{s}\lambda^{-3/2+\delta}d\lambda+\epsilon(s/t)^{3\delta+1/2}\int_{s_0}^{s}\lambda^{-3/2+2\delta}\,d\lambda\Big)\\
&\le C(C_1\varepsilon)(s/t)^{2-3\delta}+CC_1\varepsilon(s/t)^{3-4\delta}X_n(s)+CC_1\varepsilon^2(s/t)^{7/2-2\delta},
\end{aligned}
\tag{13.7}
$$

where we used that $X_n(\cdot)$ is non-decreasing and $s_0 \simeq \frac{t}{s}$. Also, recall that (7.10) gives the desired bound for $h'_{t,x}$ and, therefore, by Proposition 3.5 we deduce that

$$
(s/t)^{3\delta-2}s^{3/2}\left|\partial^I\phi\right|+(s/t)^{3-3\delta}s^{3/2}\left|\underline{\partial}_\perp\partial^I\phi\right|\le CC_0\,\varepsilon+CC_1\varepsilon+CC_1\varepsilon X_n(s).
$$

Taking the sup-norm of the above inequality in $\mathcal{K}_{[2,s]}$, we obtain $X_n(s)\le CC_0\,\varepsilon+CC_1\varepsilon+CC_1\varepsilon X_n(s)$. Then, if we take in the bootstrap assumption that ε_0' sufficiently small so that $CC_1\varepsilon\le 1/2$ for $0\le\varepsilon\le\varepsilon_0'$, we have $X_n(s)\le CC_0\,\varepsilon+CC_1\varepsilon\le CC_1\varepsilon$, which is the desired result (since $C_1\ge C_0$).

It remains to derive (13.6) and, with the notation above, we write at any (t,x)

$$
\begin{aligned}
\left|\partial^{I'}\partial_t\partial_t\phi\right| &= \left|(t/s)^2\big(\underline{\partial}_\perp-(x^a/t)\underline{\partial}_a\big)\partial^{I'}\partial_t\phi\right|\\
&\le (t/s)^2\left|\underline{\partial}_\perp\partial^{I'}\partial_t\phi\right|+(t/s)^2\left|(x^a/t)\underline{\partial}_a\partial^{I'}\partial_t\phi\right|\\
&\le (s/t)^{1-3\delta}s^{-3/2}X_n(s)+(t/s)^2t^{-1}\sum_a\left|L_a\partial^{I'}\partial_t\phi\right|,
\end{aligned}
$$

in which we used the definition of X_n and, on the other hand, the fact that $\partial^{I'}$ is of order $|I|-1$ at most. Recalling (5.16b) (together with the commutator estimates), we obtain

$$
\sum_a\left|L_a\partial^{I'}\partial_t\phi\right|\le CC_1\epsilon t^{-5/2}s^{1/2+\delta}=CC_1\epsilon(s/t)^{5/2}s^{-2+\delta},
$$

which leads us to

$$
\left|\partial^{I'}\partial_t\partial_t\phi\right|\le(s/t)^{1-3\delta}s^{-3/2}X_n(s)+t^{-1}CC_1\epsilon(s/t)^{1/2}s^{-2+\delta}.
$$

□

Before we can proceed with the proof of Proposition 13.1 in the case $|J| \geq 1$, we need to establish the following result.

Lemma 13.3. *For $|I| + |J| \leq N - 4$, one has*

$$\begin{aligned}
&\left|\partial^I L^J \left(\partial_\alpha \phi \partial_\beta \phi\right)\right| + \left|\partial^I L^J \left(\phi^2\right)\right| \\
&\leq CC_1\varepsilon (s/t)^{2-3\delta} s^{-3/2} \sum_{\substack{|I'| \leq |I| \\ \gamma}} \left|\partial^{I'} L^J \partial_\gamma \phi\right| + \left|\partial^{I'} L^J \phi\right| \\
&\quad + CC_1\varepsilon (s/t) s^{2-3\delta} s^{-3/2 + C(C_1\varepsilon)^{1/2}} \sum_{\substack{|I'| \leq |I|, |J'| < |J| \\ \gamma}} \left|\partial^{I'} L^{J'} \partial_\gamma \phi\right| + \left|\partial^{I'} L^{J'} \phi\right|.
\end{aligned} \tag{13.8}$$

Proof. We only consider $\partial_\alpha \phi \partial_\beta \phi$, by relying on (13.5) in the case $|J| = 0$. Observe that

$$\left|\partial^I L^J \left(\partial_\alpha \phi \partial_\beta \phi\right)\right| \leq \sum_{\substack{I_1 + I_2 = I \\ J_1 + J_2 = J}} \left|\partial^{I_1} L^{J_1} \partial_\alpha \phi\right| \left|\partial^{I_2} L^{J_2} \partial_\beta \phi\right|.$$

When $J_1 = 0$ or $J_2 = 0$, thanks to (11.15),

$$\left|\partial^{I_1} L^{J_1} \partial_\alpha \phi\right| \left|\partial^{I_2} L^{J_2} \partial_\beta \phi\right| \leq CC_1\varepsilon (s/t)^{2-3\delta} s^{-3/2} \sum_\gamma \left|\partial^I L^J \partial_\gamma \phi\right|.$$

When $1 \leq |J_1|$ or $1 \leq |J_2|$ we see that $|J_2| < |J|$ and $|J_1| < |J|$ and it remains to apply (11.18). □

Proof of Proposition 13.1 in the case $|J| \geq 1$. We proceed by induction and with the help of a secondary bootstrap argument (as in the proof of Proposition 11.2). We will not rewrite the argument in full details, but only provide the key steps. Suppose that on the interval $[2, s^*]$ there exist positive constants $K_{m-1}, C_{m-1}, \varepsilon'_{m-1}$ (depending only on the structure of the main system and N) such that

$$(s/t)^{3\delta-2} s^{3/2} \left|\partial^I L^J \phi\right| + (s/t)^{3\delta-3} s^{3/2} \left|\underline{\partial}_\perp \partial^I L^J \phi\right| \leq K_{m-1} C_1 \varepsilon s^{C_{m-1}(C_1\varepsilon)^{1/2}}, \tag{13.9}$$

$$t \left|L^J h_{\alpha\beta}\right| \leq K_{m-1} C_1 \varepsilon s^{C_{m-1}(C_1\varepsilon)^{1/2}} \tag{13.10}$$

for $0 \leq \varepsilon \leq \varepsilon'_{m-1}$ and $|I| + |J| \leq N - 4$ and $|J| \leq m - 1 < N - 4$. We will prove that there exist positive constants K_m, C_m, ε'_m (determined by the structure of the main system and the integer N) such that the following inequalities hold for $0 \leq \varepsilon \leq \varepsilon'_m$:

$$(s/t)^{3\delta-2} s^{3/2} \left|\partial^I L^J \phi\right| + (s/t)^{3\delta-3} s^{3/2} \left|\underline{\partial}_\perp \partial^I L^J \phi\right| \leq K_m C_1 \varepsilon s^{C_m(C_1\varepsilon)^{1/2}}, \tag{13.11}$$

$$t\left|L^J h_{\alpha\beta}\right| \leq K_m C_1 \varepsilon s^{C_m (C_1\varepsilon)^{1/2}}. \tag{13.12}$$

We begin the formulation of the secondary bootstrap argument and set

$$s^{**} := \sup_{s\in[2,s^*]} \{s|(13.11) \text{ and } (13.12) \text{ hold in } \mathcal{K}_{[2,s^*]}\}.$$

Suppose the K_m that we have taken is sufficiently large such that $s^{**} > 2$ and $C_m = 2C_{m-1}$ (see the argument in the proof of Proposition 11.2.)

We substitute the assumptions (13.9), (13.10), (13.11) and (13.12) into (13.2). This gives

$$\begin{aligned}&\left|[\partial^I L^J, h^{\mu\nu}\partial_\mu\partial_\nu]\phi\right| \\ &\leq C(C_1\varepsilon)^2(s/t)^3 s^{-3+3\delta} + CK_m^2(C_1\varepsilon)^2(s/t)^{2-3\delta} s^{\;5/2+C_m(C_1\varepsilon)^{1/2}}.\end{aligned} \tag{13.13}$$

With the notation in Proposition 3.5 (recalling that $h'_{t,x}$ is bounded in view of (7.10) and R_i are bounded by Lemma 11.3), we obtain

$$|F(s)| \leq CC_1\varepsilon(s/t)^{3/2} s_0^{-1/2+3\delta} + CC_m^{-1}K_m^2(C_1\varepsilon)^{3/2}(s/t)^{2-3\delta} s^{C_m(C_1\varepsilon)^{1/2}}.$$

Then in view of (3.5), we have

$$\begin{aligned}&(s/t)^{3\delta-2}s^{3/2}\left|\partial^I L^J\phi\right| + (s/t)^{3\delta-3}s^{3/2}\left|\underline{\partial}_\perp\partial^I L^J\phi\right| \\ &\leq CK_{0,m}C_1\varepsilon + CC_1\varepsilon + CC_m^{-1}K_m^2(C_1\varepsilon)^{3/2}s^{C_m(C_1\varepsilon)^{1/2}}.\end{aligned}$$

Then, as in the proof of Proposition 11.2, we choose

$$\varepsilon'_m = \frac{C_m^2}{C_1}\left(\frac{K_m - 2CK_{0,m} - 2C}{2CK_m^2}\right)^2.$$

Then, for $0 \leq \varepsilon \leq \varepsilon'_m$, we have

$$(s/t)^{3\delta-2}s^{3/2}\left|\partial^I L^J\phi\right| + (s/t)^{3\delta-3}s^{3/2}\left|\underline{\partial}_\perp\partial^I L^J\phi\right| \leq \frac{1}{2}K_m C_1\varepsilon s^{C(C_1\varepsilon)^{1/2}}.$$

The estimate for $L^J h_{\alpha\beta}$ is checked as the argument in the proof of Proposition 11.2. We omit the details and point out the estimates on QS_ϕ is covered by Lemma 13.3 and the induction-bootstrap assumption (13.9), (13.10), (13.11) and (13.12). Other nonlinear terms such as $F_{\alpha\beta}$ and $h^{\mu\nu}\partial_\mu\partial_\nu h_{\alpha\beta}$ are bounded in view of (12.21) and (13.1). The same argument as in the proof of Proposition 11.2 leads us to the desired result with $\varepsilon_4 = \min(\varepsilon'_m, \varepsilon'_0)$, where ε'_0 was determined at the end of the proof for $|J| = 0$. □

Chapter 14

Low-Order Refined Energy Estimate for the Scalar Field

It remains to establish the refined energy estimate in order to complete the proof of our main result.

Proposition 14.1. *Let $|I| + |J| \leq N - 4$ and suppose that the bootstrap assumptions* (5.1) (5.2) *hold for C_1 sufficiently large, then there exists some $\varepsilon_5 > 0$ such that for all $0 \leq \varepsilon \leq \varepsilon_5$;*

$$E_{M,c^2}(s, \partial^I L^J \phi)^{1/2} \leq \frac{1}{2} C_1 \varepsilon s^{C(C_1\varepsilon)^{1/2}}. \tag{14.1}$$

Proof. Our argument now relies on the energy estimate in Proposition 3.2, in which the coercivity condition (3.1) is guaranteed by Lemma 7.2. The estimate for $M[\partial^I L^J \phi]$ is provided by (7.7b). So the only issue still to be discussed is the estimate of the commutator $\big\|[\partial^I L^J, h^{\mu\nu}\partial_\mu\partial_\nu]\phi\big\|_{L^2(\mathcal{H}_s^*)}$. Here, we use (4.16) and, in view of (6.8), obtain

$$\|GQQ_{h\phi}(N-4,k)\|_{L^2_f(\mathcal{H}_s)} \leq C(C_1\varepsilon)^2 s^{-3/2+2\delta}.$$

For $t^{-1}\partial^{I_3}L^{J_3}h_{\alpha'\beta'}\partial^{I_4}L^{J_4}\partial_\gamma\phi$, we have

$$\begin{aligned}
&\|t^{-1}\partial^{I_3}L^{J_3}h_{\alpha'\beta'}\partial^{I_4}L^{J_4}\partial_\gamma\phi\|_{L^2_f(\mathcal{H}_s)} \\
&\leq \Big\|t^{-1}(t^{-1} + (s/t)t^{-1/2}s^\delta)\partial^{I_4}L^{J_4}\partial_\gamma\phi\Big\|_{L^2_f(\mathcal{H}_s)} \leq C(C_1\varepsilon)^2 s^{-3/2+2\delta},
\end{aligned}$$

while the term $\partial^{I_1}L^{J_1}\underline{h}^{00}\partial^{I_2}L^{J_2}\partial_t\partial_t\phi$ is bounded by applying (7.1):

$$\begin{aligned}
&\|\partial^{I_1}L^{J_1}\underline{h}^{00}\partial^{I_2}L^{J_2}\partial_t\partial_t\phi\|_{L^2(\mathcal{H}_s^*)} \\
&\leq CC_1\varepsilon s^{-3/2+\delta}\|(s/t)^{3/2}\partial^{I_2}L^{J_2}\partial_t\partial_t\phi\|_{L^2(\mathcal{H}_s^*)} \leq CC_1\varepsilon s^{-3/2+2\delta}.
\end{aligned}$$

The term $L^{J_1'}\underline{h}^{00}\partial^I L^{J_2'}\partial_t\partial_t\phi$ is bounded by applying (13.4) and observing that $|J_1'| > 0$:

$$
\begin{aligned}
\|L^{J_1'}\underline{h}^{00}\partial^I L^{J_2'}\partial_t\partial_t\phi\|_{L^2_f(\mathcal{H}_s)} &\leq CC_1\varepsilon\|t^{-1}s^{C(C_1\varepsilon)^{1/2}}\partial^I L^{J_2'}\partial_t\partial_t\phi\|_{L^2_f(\mathcal{H}_s)} \\
&\leq CC_1\varepsilon s^{-1+C(C_1\varepsilon)^{1/2}}\|(s/t)\partial^I L^{J_2'}\partial_t\partial_t\phi\|_{L^2_f(\mathcal{H}_s)} \\
&\leq CC_1\varepsilon s^{-1+C(C_1\varepsilon)^{1/2}}\sum_{|J'|<|J|} E_{M,c^2}(s,\partial^I L^{J'}\phi)^{1/2}.
\end{aligned}
$$

And for the term $\underline{h}^{00}\partial_\alpha\partial_\beta$, we apply (11.16):

$$
\|\underline{h}^{00}\partial_\alpha\partial_\beta\partial^I L^{J'}\|_{L^2_f(\mathcal{H}_s)} \leq CC_1\varepsilon s^{-1}\sum_{|J'|<|J|} E_{M,c^2}(\partial^I L^{J'}\phi)^{1/2},
$$

so that

$$
\|[\partial^I L^J, h^{\mu\nu}\partial_\mu\partial_\nu]\phi\|_{L^2(\mathcal{H}_s^*)} \leq CC_1\varepsilon s^{-1+C(C_1\varepsilon)^{1/2}}\sum_{|J'|<|J|} E_{M,c^2}(s,\partial^I L^{J'}\phi)^{1/2}.
$$

So by Proposition 3.2, we have

$$
\begin{aligned}
E_{M,c^2}(s,\partial^I L^J\phi)^{1/2} &\leq C_0\,\varepsilon + C(C_1\varepsilon)^2\int_2^s \tau^{-3/2+2\delta}d\tau \\
&+ CC_1\varepsilon\sum_{|J'|<|J|}\int_2^s \tau^{-1+C(C_1\varepsilon)^{1/2}}E_{M,c^2}(\tau,\partial^I L^{J'}\phi)^{1/2}d\tau.
\end{aligned}
\tag{14.2}
$$

When $|J| = 0$, the last term disappears. We have

$$
E_{M,c^2}(s,\partial^I\phi)^{1/2} \leq CC_0\,\varepsilon + C(C_1\varepsilon)^2. \tag{14.3}
$$

We are going to prove that for all $|I| + |J| \leq N-4$,

$$
E_{M,c^2}(s,\partial^I L^J\phi)^{1/2} \leq CC_0\,\varepsilon + C(C_1\varepsilon)^{3/2}s^{C(C_1\varepsilon)^{1/2}}. \tag{14.4}
$$

When $|J| \geq 1$, we proceed by induction on $|J|$ and see that (14.4) is guaranteed by (14.3) ($C_1\varepsilon$ smaller that 1). Assume that (14.4) holds for $|J| \leq m-1 < n-4$, we will prove it for $|J| = m \leq N-4$. We directly apply the induction assumption in (14.2) and conclude that

$$
E_{M,c^2}(s,\partial^I L^J\phi)^{1/2} \leq CC_0\,\varepsilon + C(C_1\varepsilon)^{3/2}s^{C(C_1\varepsilon)^{1/2}}
$$

for $|I| + |J| \leq N-4$ and, by taking

$$
\varepsilon_5 = \left(\frac{C_1 - 2CC_0}{2CC_1^{3/2}}\right)^2,
$$

the desired result is proven. □

In conclusion, in view of (10.5), (12.7), (12.8) and (14.1), if the bootstrap assumption holds for $C_1 > C_0$ sufficiently large, then there exists some $\varepsilon_0 := \min\{\varepsilon_1\,\varepsilon_2, \varepsilon_3, \varepsilon_4, \varepsilon_5\}$ such that

$$
\begin{aligned}
E_M(s, \partial^I L^J h_{\alpha\beta})^{1/2} &\leq \frac{1}{2} C_1 \varepsilon s^{C(C_1\varepsilon)^{1/2}}, && |I| + |J| \leq N, \\
E_M(s, \partial^I L^J \phi)^{1/2} &\leq \frac{1}{2} C_1 \varepsilon s^{1/2 + C(C_1\varepsilon)^{1/2}}, && N - 3 \leq |I| + |J| \leq N, \\
E_M(s, \partial^I L^J \phi)^{1/2} &\leq \frac{1}{2} C_1 \varepsilon s^{C(C_1\varepsilon)^{1/2}}, && |I| + |J| \leq N - 4.
\end{aligned}
$$

This improves the bootstrap assumption (5.1)–(5.2). We see that (5.1)–(5.2) hold on the time interval where the solution exists. In view of the local existence theory for the hyperboloidal foliation (see the last chapter of [39]) the global existence result is thus established.

Appendix A

Revisiting the wave-Klein-Gordon model

A *wave-Klein-Gordon model* was "extracted" from the Einstein equations by the authors in [40; 41] when they were beginning to analyze the Einstein equations via the Hyperboloidal Foliation Method introduced in [39]. This model[1] provided to the authors a simple, yet highly not trivial, example of coupling between a wave equation and a Klein-Gordon equation, before developing the method for the full Einstein system, as we do in the present monograph. We revisit here the proof of existence in [41] since our presentation missed one bootstrap condition in the list (5.1) which however turns out to be necessary for dealing with the (comparatively easier) wave component when $k = 0$ in (5.1).

When $k = 0$, the first bound in (5.1) in [41] should be weakened to

$$E_m(s, \partial^I u)^{1/2} \leq C_1 \varepsilon s^\delta, \qquad |I| \leq N, \tag{A.1}$$

while a similar remark applies to (5.2). Doing so has no effect on the derivation of the sup-norm bounds (in Section 6.2, on which Section 7 is based), since in the application of the Klainerman-Sobolev inequality one uses one boost at least, and the additional growth allowed by (A.1) is negligible. Note in passing also that, in Section 6.5 of [41], the Hardy-based estimate (6.20a) is valid for $k = |J| \geq 1$ only, while we already pointed out in [41] the next inequality (6.20b) is never used.

In Lemma 8.1, the estimate (8.4) can be improved to

$$M(s) \lesssim C_1 \varepsilon s^{-3/2+k\delta}, \tag{A.2}$$

[1] A.D. Ionescu and B. Pausader recently further investigated our model via Fourier techniques; see ArXiv:1703.02846.

which is checked for $|I|+|J|\leq N-1$ by writing

$$\int_{\mathcal{H}_s}\left|\partial_\gamma h^{\alpha\beta}\partial_\alpha\partial^I L^J v\right|^2 dx \lesssim C_1\varepsilon\int_{\mathcal{H}_s} t^{-1}s^{-2}\left|\partial_\alpha\partial^I L^J v\right|^2 dx$$
$$\lesssim C_1\varepsilon s^{-3}\int_{\mathcal{H}_s}\left|\partial_\alpha\partial^I L^J v\right|^2 dx \lesssim C_1\varepsilon s^{-3}\sum_\alpha E_{g,c}(s,\partial_\alpha\partial^I L^J v).$$

In Lemma 8.2, when $k=0$ (8.6) can be improved to

$$\|[H^{\alpha\beta}u\partial_\alpha\partial_\beta,\partial^I]v\|_{L^2_f(\mathcal{H}_s)}\lesssim (C_1\varepsilon)^2 s^{-3/2+2\delta},\quad |I|\leq N. \tag{A.3}$$

Namely, only the term $\partial^{I_1}L^{J_1}u\partial^{I_2}L^{J_2}\partial_\alpha\partial_\beta v$ with $|I_1|=1$ and $J_1=0$ need to be considered:

$$\begin{aligned}\|\partial_\gamma u\partial^{I_2}L^J\partial_\alpha\partial_\beta v\|_{L^2_f(\mathcal{H}_s)} \leq&\|(s/t)\partial_\gamma u\|_{L^2(\mathcal{H}_s)}\,\|(t/s)\partial^{I_2}L^J\partial_\alpha\partial_\beta v\|_{L^\infty(\mathcal{H}_s)}\\ \lesssim&(C_1\varepsilon)^2\|(s/t)^{-\delta}s^{-3/2}\|_{L^\infty(\mathcal{H}_s)}\lesssim (C_1\varepsilon)^2 s^{-3/2+\delta}.\end{aligned} \tag{A.4}$$

In Lemma 8.3, when $k=0$ (8.7) can be improved to

$$\left\|\partial^I\left(P^{\alpha\beta}\partial_\alpha v\partial_\beta v+Rv^2\right)\right\|\lesssim (C_1\varepsilon)^2 s^{-1+\delta},\qquad |I|\leq N. \tag{A.5}$$

Namely, in $\partial^I(\partial_\alpha v\partial_\beta v)=\sum_{I_1+I_2=I}\partial^{I_1}\partial_\alpha v\,\partial^{I_2}\partial_\beta v$ we can assume that $|I_1|\leq|I_2|$, hence $|I_1|\leq[|I|/2]\leq N-5$, and then by (7.23b) (with $\partial^I\partial_\alpha$ of order $\leq N-4$) and (6.5) (third and last inequalities):

$$\begin{aligned}&\left\|\partial^I\left(\partial_\alpha v\partial_\beta v\right)\right\|_{L^2_f(\mathcal{H}_s)}\lesssim (C_1\varepsilon)^2\|(s/t)^{1/2-4\delta}t^{-3/2}(t/s)\;(s/t)\partial^{I_2}\partial_\beta v\|_{L^2_f(\mathcal{H}_s)}\\&\lesssim (C_1\varepsilon)^2 s^{-1+\delta}.\end{aligned} \tag{A.6}$$

In the proof of Proposition 5.1, when $|J|=0$ thanks to (A.5)

$$E_m(s,\partial^I u)^{1/2}\leq\overline{C}C_0\varepsilon+\overline{C}(C_1\varepsilon)^2\int_2^s\overline{s}^{-1+\delta}\,d\overline{s}\leq\overline{C}C_0\varepsilon+\overline{C}(C_1\varepsilon)^2 s^\delta, \tag{A.7}$$

and for (8.14) with $k=0$, one has $|I|\leq N-4$ and we can apply (A.2)-(A.3):

$$E_{m,c}(s,\partial^I v)^{1/2}\leq\overline{C}C_0\varepsilon+\overline{C}(C_1\varepsilon)^2\int_2^s\overline{s}^{-3/2+k\delta}\,d\overline{s}\leq\overline{C}C_0\varepsilon+\overline{C}(C_1\varepsilon)^2. \tag{A.8}$$

Appendix B

Sup-norm estimate for the wave equations

Proposition B.1. *Let u be a spatially compactly supported solution to the wave equation*

$$
\begin{aligned}
-\Box u &= f, \\
u|_{t=2} = 0, \qquad & \partial_t u|_{t=2} = 0,
\end{aligned}
\tag{B.1}
$$

in which f is spatially compactly supported in $\mathcal{K}$ and satisfies

$$
|f| \leq C_f t^{-2-\nu}(t-r)^{-1+\mu} \tag{B.2}
$$

for some $C_f > 0$, $0 < \mu \leq 1/2$, and $0 < |\nu| \leq 1/2$. Then, one has

$$
|u(t,x)| \lesssim \begin{cases} \frac{C_f}{\nu\mu}(t-r)^{\mu-\nu}t^{-1}, & 0 < \nu \leq 1/2, \\ \frac{C_f}{|\nu|\mu}(t-r)^{\mu}t^{-1-\nu}, & -1/2 \leq \nu < 0. \end{cases} \tag{B.3}
$$

We denote by $d\sigma$ the Lebesgue measure on the sphere $\{|y| = 1-\lambda\}$ and $x \in \mathbb{R}^3$ with $r = |x|$, and consider the integral term

$$
I(\lambda) = I(\lambda, t, x/t) := \int_{|y|=1-\lambda, |\frac{x}{t}-y| \leq \lambda - t^{-1}} \frac{d\sigma(y)}{\left(\lambda - |\frac{x}{t} - y|\right)^{1-\mu}}.
$$

Clearly, when $0 < \lambda \leq \frac{t-r+1}{2t}$, we have $I(\lambda) = 0$.

Lemma B.1. *When $\frac{t-r+1}{2t} \leq \lambda \leq 1$, we obtain*

$$
I(\lambda) \lesssim \begin{cases} \dfrac{\lambda t(1-\lambda)}{\mu r}\left(\dfrac{t-r}{t}\right)^{\mu}, & \dfrac{t-r+1}{2t} \leq \lambda \leq \dfrac{t+r+1}{2t}, \\ (1-\lambda)\left(\dfrac{t+r}{t} - \lambda\right)\left(2\lambda - \dfrac{t+r}{t}\right)^{-1+\mu}, & \dfrac{t+r+1}{2t} \leq \lambda \leq \dfrac{t-r}{t}, \\ & \textit{provided } \dfrac{t+r+1}{2t} \leq \dfrac{t-r}{t}, \\ \dfrac{(1-\lambda)t}{\mu r}\left(\dfrac{t-r}{t}\right)^{\mu}, & \max\left(\dfrac{t-r}{t}, \dfrac{t+r+1}{2t}\right) \leq \lambda \leq 1. \end{cases}
$$

Proof of Proposition B.1. From the expression

$$u(t,x) = \frac{1}{4\pi}\int_2^t \frac{1}{t-\overline{s}}\int_{|y|=t-\overline{s}} f(\overline{s}, x-y)\,d\sigma d\overline{s}, \tag{B.4}$$

in which the integration is on the intersection of the cone $\{(\overline{s},y)\,/\,|y-x| = t-\overline{s}, 2 \leq \overline{s} \leq t\}$ and $\{(t,x)\,/\,r < t-1, t^2-r^2 \leq s^2, t \geq 2\}$, we obtain

$$\begin{aligned}
|u(t,x)| &\leq \frac{C_f}{4\pi}\int_2^t \int_{|y|=t-\overline{s},|x-y|\leq \overline{s}-1} \frac{\overline{s}^{-2-\nu}(\overline{s}-|x-y|)^{-1+\mu}}{t-\overline{s}} d\sigma d\overline{s} \\
&= \frac{C_f}{4\pi t^{1+\nu-\mu}}\int_{\frac{2}{t}}^1 \int_{|y'|=1-\lambda,|\frac{x}{t}-y'|\leq \lambda - t^{-1}} \frac{(1-\lambda)^{-1}\lambda^{-2-\nu}d\sigma d\lambda}{\left(\lambda - |\frac{x}{t}-y'|\right)^{1-\mu}} \\
&\qquad\qquad (\lambda := \overline{s}/t, \quad y' := y/t) \\
&= \frac{C_f}{4\pi t^{1+\nu-\mu}}\int_{\frac{2}{t}}^1 (1-\lambda)^{-1}\lambda^{-2-\nu} \int_{|y'|=1-\lambda,|\frac{x}{t}-y'|\leq \lambda-t^{-1}} \frac{d\sigma}{\left(\lambda - |\frac{x}{t}-y'|\right)^{1-\mu}}\, d\lambda.
\end{aligned}$$

When $|\frac{x}{t}-y'| \leq \lambda - t^{-1}$, we get $\frac{t-r+1}{2t} \leq \lambda \leq 1$. In the following, we replace y' by y. We distinguish between two cases:

Case 1: $\frac{t-r}{t} > \frac{t+r+1}{2t} \Leftrightarrow r \leq \frac{t-1}{3}$. We write

$$\begin{aligned}
&|u(t,x)| \\
&\leq \frac{C_f}{4\pi t^{1+\nu-\mu}}\int_{\frac{t-r+1}{2t}}^1 (1-\lambda)^{-1}\lambda^{-2-\nu}\int_{|y|=1-\lambda,|\frac{x}{t}-y|\leq\lambda-t^{-1}} \frac{d\sigma}{\left(\lambda-|\frac{x}{t}-y|\right)^{1-\mu}}\,d\lambda \\
&\lesssim \frac{C_f}{\mu t^{1+\nu-\mu}}\int_{\frac{t-r+1}{2t}}^{\frac{t+r+1}{2t}} (1-\lambda)^{-1}\lambda^{-2-\nu}\frac{\lambda t(1-\lambda)}{r}\left(\frac{t-r}{t}\right)^{\mu} d\lambda \\
&\quad + \frac{C_f}{t^{1+\nu-\mu}}\int_{\frac{t+r+1}{2t}}^{\frac{t-r}{t}} (1-\lambda)^{-1}\lambda^{-2-\nu}(1-\lambda)\left(\frac{t+r}{t}-\lambda\right)\left(2\lambda - \frac{t+r}{t}\right)^{-1+\mu} d\lambda \\
&\quad + \frac{C_f}{\mu t^{1+\nu-\mu}}\int_{\frac{t-r}{t}}^1 (1-\lambda)^{-1}\lambda^{-2-\nu}\frac{(1-\lambda)t}{r}\left(\frac{t-r}{t}\right)^{\mu} d\lambda,
\end{aligned}$$

and therefore

$$\begin{aligned}
|u(t,x)| \lesssim& \frac{C_f}{\mu t^{1+\nu-\mu}}\frac{t}{r}\left(\frac{t-r}{t}\right)^{\mu}\int_{\frac{t-r+1}{2t}}^{\frac{t+r+1}{2t}} \lambda^{-1-\nu}\,d\lambda \\
&+ \frac{C_f}{t^{1+\nu-\mu}}\int_{\frac{t+r+1}{2t}}^{\frac{t-r}{t}} \lambda^{-2-\nu}\left(\frac{t+r}{t}-\lambda\right)\left(2\lambda-\frac{t+r}{t}\right)^{-1+\mu} d\lambda \\
&+ \frac{C_f}{\mu t^{1+\nu-\mu}}\frac{t}{r}\left(\frac{t-r}{t}\right)^{\mu}\int_{\frac{t-r}{t}}^1 \lambda^{-2-\nu}\,d\lambda.
\end{aligned}$$

Recall that $r \le \frac{t-1}{3}$ and that $0 < |\nu| \le 1/2$, we have

$$\frac{t}{r}\int_{\frac{t-r+1}{2t}}^{\frac{t+r+1}{2t}} \lambda^{-1-\nu}\, d\lambda \lesssim \left(\frac{t}{t-r}\right)^{1+\nu} \lesssim 1,$$

and

$$\left|\frac{C_f}{\mu t^{1+\nu-\mu}}\frac{t}{r}\left(\frac{t-r}{t}\right)^{\mu}\int_{\frac{t-r+1}{2t}}^{\frac{t+r+1}{2t}} \lambda^{-1-\nu}\, d\lambda\right| \lesssim C_f\mu^{-1}(t-r)^{\mu}t^{-1-\nu}.$$

For the second integral term, we note that

$$\begin{aligned}&\int_{\frac{t+r+1}{2t}}^{\frac{t-r}{t}} \lambda^{-2-\nu}\left(\frac{t+r}{t}-\lambda\right)\left(2\lambda-\frac{t+r}{t}\right)^{-1+\mu} d\lambda \\ &\lesssim \int_{\frac{t+r+1}{2t}}^{\frac{t-r}{t}} \left(2\lambda-\frac{t+r}{t}\right)^{-1+\mu} d\lambda = \frac{1}{\mu}\left(2\lambda-\frac{t+r}{t}\right)^{\mu}\Bigg|_{\frac{t+r+1}{2t}}^{\frac{t-r}{t}} \lesssim \frac{1}{\mu},\end{aligned}$$

thus

$$\frac{C_f}{t^{1+\nu-\mu}}\int_{\frac{t+r+1}{2t}}^{\frac{t-r}{t}} \lambda^{-2-\nu}\left(\frac{t+r}{t}-\lambda\right)\left(2\lambda-\frac{t+r}{t}\right)^{-1+\mu} d\lambda \lesssim \frac{C_f}{\mu t^{1+\nu-\mu}}.$$

For the third term, from $\frac{t-r}{t} \ge \frac{t+r+t1}{2t} \ge \frac{1}{2}$ we get

$$\begin{aligned}\frac{C_f}{\mu t^{1+\nu-\mu}}\frac{t}{r}\left(\frac{t-r}{t}\right)^{\mu}\int_{\frac{t-r}{t}}^{1} \lambda^{-2-\nu}\, d\lambda \lesssim &\frac{C_f}{\mu t^{1+\nu-\mu}}\frac{t}{r}\left(\frac{t-r}{t}\right)^{\mu}\int_{\frac{t-r}{t}}^{1} 2^{2+\mu}\, d\lambda \\ \lesssim &C_f\mu^{-1}(t-r)^{\mu}t^{-1-\nu}.\end{aligned}$$

Hence, in the case $0 < r \le \frac{t-1}{3}$, $|u(t,x)| \lesssim C_f\mu^{-1}(t-r)^{\mu}t^{-1-\nu}$.

Case 2: $\frac{t+r+1}{2t} \ge \frac{t-r}{t} \Leftrightarrow r \ge \frac{t-1}{3}$. The second case in Lemma B.1 can not occur. We have

$$|u(t,x)| \lesssim \frac{C_f}{\mu t^{1+\nu-\mu}}\left(\frac{t-r}{t}\right)^{\mu}\left(\int_{\frac{t-r+1}{2t}}^{\frac{t+r+1}{2t}} \lambda^{-1-\nu}\, d\lambda + \int_{\frac{t+r+1}{2t}}^{1} \lambda^{-2-\nu}\, d\lambda\right).$$

Since $\frac{t+r+1}{2t} \ge 1/2$, the second integral term is bounded by some constant C. For the first integral, when $\nu > 0$,

$$\int_{\frac{t-r+1}{2t}}^{\frac{t+r+1}{2t}} \lambda^{-1-\nu}\, d\lambda \lesssim \frac{1}{\nu}\left(\frac{t-r+1}{t}\right)^{-\nu},$$

thus $|u(t,x)| \lesssim C_f(\mu\nu)^{-1}(t-r)^{\mu-\nu}t^{-1}$.

When $\nu < 0$, we write

$$\int_{\frac{t-r+1}{2t}}^{\frac{t+r+1}{2t}} \lambda^{-1-\nu}\, d\lambda \lesssim \frac{1}{|\nu|}\left(\frac{t+r+1}{t}\right)^{-\nu} \lesssim \frac{1}{|\nu|}$$

and obtain $|u(t,x)| \lesssim C_f(\mu|\nu|)^{-1}(t-r)^{\mu}t^{-1-\nu}$. □

Proof of Lemma B.1. When $r = 0$, the estimate is trivial. When $r > 0$, we can set $x = (r, 0, 0)$. The surface $S_\lambda := \{|y| = 1-\lambda\} \cap \{|\frac{x}{t} - y| \leq \lambda - t^{-1}\}$ is parameterized by:

- θ: angle from $(1,0,0)$ to y with $0 \leq \theta \leq \pi$,
- ϕ: angle from the plane determined by $(1,0,0)$ and $(0,1,0)$ and the plane determined by y and $(1,0,0)$ with $0 \leq \phi \leq 2\pi$.

We have $y = (1-\lambda)\big(\cos\theta, \sin\theta\cos\phi, \sin\theta\sin\phi\big)$ and distinguish between two cases:

Case 1. When $\frac{t-r+1}{2t} \leq \lambda \leq \frac{t+r+1}{2t}$, we only have a part of the sphere $\{|y| = 1-\lambda\}$ contained in the ball $\{|\frac{x}{t} - y| \leq \lambda - t^{-1}\}$ where $\cos(\theta) \geq \frac{(r/t)^2 + (1-\lambda)^2 - (\lambda - t^{-1})^2}{(2r/t)(1-\lambda)}$. So we set $\theta_0 := \arccos\left(\frac{(r/t)^2 + (1-\lambda)^2 - (\lambda - t^{-1})^2}{(2r/t)(1-\lambda)}\right)$ and see that

$$\lambda - \left|\frac{x}{t} - y\right| = \lambda - \sqrt{\frac{r^2}{t^2} + (1-\lambda)^2 - 2\frac{r}{t}(1-\lambda)\cos\theta}$$

and $d\sigma = (1-\lambda)^2 \sin(\theta) d\theta d\phi$. The integral is estimated as follows:

$$\begin{aligned}
&\int_{|y|=1-\lambda, |\frac{x}{t}-y| \leq \lambda - t^{-1}} \frac{d\sigma}{\big(\lambda - |\frac{x}{t} - y|\big)^{1-\mu}} \\
&= \int_0^{2\pi} d\phi \int_0^{\theta_0} (1-\lambda)^2 \sin\theta \left(\lambda - \sqrt{\frac{r^2}{t^2} + (1-\lambda)^2 - 2\frac{r}{t}(1-\lambda)\cos\theta}\right)^{-1+\mu} d\theta \\
&= 2\pi \int_0^{\theta_0} (1-\lambda)^2 \sin\theta \left(\lambda - \sqrt{\frac{r^2}{t^2} + (1-\lambda)^2 - 2\frac{r}{t}(1-\lambda)\cos\theta}\right)^{-1+\mu} d\theta \\
&= -2\pi(1-\lambda)^2 \int_0^{\theta_0} \left(\lambda - \sqrt{\frac{r^2}{t^2} + (1-\lambda)^2 - 2\frac{r}{t}(1-\lambda)\cos\theta}\right)^{-1+\mu} d\cos\theta
\end{aligned}$$

thus, with $\omega = \cos\theta$,

$$\begin{aligned}
&\int_{|y|=1-\lambda, |\frac{x}{t}-y| \leq \lambda - t^{-1}} \frac{d\sigma}{\big(\lambda - |\frac{x}{t} - y|\big)^{1-\mu}} \\
&= 2\pi(1-\lambda)^2 \int_{\cos\theta_0}^{1} \left(\lambda - \sqrt{\frac{r^2}{t^2} + (1-\lambda)^2 - 2\frac{r}{t}(1-\lambda)\omega}\right)^{-1+\mu} d\omega \\
&= \frac{\pi t(1-\lambda)}{r} \int_{|\frac{r}{t} - (1-\lambda)|^2}^{(\lambda - t^{-1})^2} \big(\lambda - \sqrt{\gamma}\big)^{-1+\mu} d\gamma \\
&= 2\frac{\pi t(1-\lambda)}{r} \int_{t^{-1}}^{\lambda - |\frac{r}{t} - (1-\lambda)|} \zeta^{-1+\mu}(\lambda - \zeta)\, d\zeta,
\end{aligned}$$

where $\gamma = \frac{r^2}{t^2} + (1-\lambda)^2 - 2\frac{r}{t}(1-\lambda)\omega$ and $\zeta := \lambda - \sqrt{\gamma}$. We distinguish between two sub-cases.

Case 1.1: $\frac{r}{t} \leq 1-\lambda$ or, equivalently, $\lambda \leq \frac{t-r}{t}$. We have

$$\begin{aligned}
&2\frac{\pi t(1-\lambda)}{r}\int_{t^{-1}}^{\lambda-|\frac{r}{t}-(1-\lambda)|}\zeta^{-1+\mu}(\lambda-\zeta)\,d\zeta \\
&= 2\frac{\pi t(1-\lambda)}{r}\int_{t^{-1}}^{2(\lambda-\frac{t-r}{2t})}\zeta^{-1+\mu}(\lambda-\zeta)\,d\zeta \lesssim \frac{\lambda t(1-\lambda)}{\mu r}\frac{(t-r)^{\mu}}{t^{\mu}}.
\end{aligned}$$

Case 1.2: $1-\lambda < \frac{r}{t}$ or, equivalently, $\lambda > \frac{t-r}{t}$. We have

$$\begin{aligned}
&2\frac{\pi t(1-\lambda)}{r}\int_{t^{-1}}^{\lambda-|\frac{r}{t}-(1-\lambda)|}\zeta^{-1+\mu}(\lambda-\zeta)\,d\zeta \\
&= 2\frac{\pi t(1-\lambda)}{r}\int_{t^{-1}}^{\frac{t-r}{t}}\zeta^{-1+\mu}(\lambda-\zeta)\,d\zeta \lesssim \frac{\lambda t(1-\lambda)}{\mu r}\frac{(t-r)^{\mu}}{t^{\mu}}.
\end{aligned}$$

Case 2. When $\frac{t+r+1}{2t} \leq \lambda \leq 1$, the sphere $\{|y| = 1-\lambda\}$ is contained in $\{|(x/t)-y| \leq \lambda - t^{-1}\}$ and

$$\begin{aligned}
&\int_{|y|=1-\lambda, |\frac{x}{t}-y|\leq\lambda-t^{-1}}\frac{d\sigma}{\left(\lambda-|\frac{x}{t}-y|\right)^{1-\mu}} = \int_{|y|=1-\lambda}\frac{d\sigma}{\left(\lambda-|\frac{x}{t}-y|\right)^{1-\mu}} \\
&= 2\pi\int_0^{\pi}(1-\lambda)^2\sin\theta\left(\lambda-\sqrt{\frac{r^2}{t^2}+(1-\lambda)^2-2\frac{r}{t}(1-\lambda)\cos\theta}\right)^{-1+\mu}d\theta \\
&= 2\pi(1-\lambda)^2\int_{-1}^{1}\left(\lambda-\sqrt{\frac{r^2}{t^2}+(1-\lambda)^2-2\frac{r}{t}(1-\lambda)\omega}\right)^{-1+\mu}d\omega.
\end{aligned}$$

Therefore, we have

$$\begin{aligned}
&\int_{|y|=1-\lambda, |\frac{x}{t}-y|\leq\lambda-t^{-1}}\frac{d\sigma}{\left(\lambda-|\frac{x}{t}-y|\right)^{1-\mu}} \\
&= 2\frac{\pi t(1-\lambda)}{r}\int_{\lambda-(\frac{r}{t}+(1-\lambda))}^{\lambda-|\frac{r}{t}-(1-\lambda)|}\zeta^{-1+\mu}(\lambda-\zeta)\,d\zeta \\
&= 2\frac{\pi t(1-\lambda)}{r}\int_{2\lambda-\frac{t+r}{t}}^{\lambda-|\frac{r}{t}-(1-\lambda)|}\zeta^{-1+\mu}(\lambda-\zeta)\,d\zeta.
\end{aligned}$$

We distinguish between two sub-cases.

Case 2.1: When $\frac{r}{t} \leq 1 - \lambda$ or, equivalently, $\lambda \leq \frac{t-r}{t}$, we have

$$
\begin{aligned}
&2\frac{\pi t(1-\lambda)}{r} \int_{2\lambda - \frac{t+r}{t}}^{\lambda - |\frac{r}{t} - (1-\lambda)|} \zeta^{-1+\mu}(\lambda - \zeta)\, d\zeta \\
&= 2\frac{\pi t(1-\lambda)}{r} \int_{2\lambda - \frac{t+r}{t}}^{2\lambda - \frac{t-r}{t}} \zeta^{-1+\mu}(\lambda - \zeta)\, d\zeta \\
&\leq C(1-\lambda)\left(\frac{t+r}{t} - \lambda\right)\left(2\lambda - \frac{t+r}{t}\right)^{-1+\mu},
\end{aligned}
$$

where the function $\zeta^{-1+\mu}(\lambda-\zeta)$ is decreasing and we can bound this integral by the value at the inferior boundary (which is $2\lambda - \frac{t+r}{t}$) times the length of the interval $2r/t$.

Case 2.2: When $1 - \lambda < \frac{r}{t}$ or, equivalently, $\lambda > \frac{t-r}{t}$, we have

$$
\begin{aligned}
&2\frac{\pi t(1-\lambda)}{r} \int_{2\lambda - \frac{t+r}{t}}^{\lambda - |\frac{r}{t} - (1-\lambda)|} \zeta^{-1+\mu}(\lambda - \zeta)\, d\zeta \\
&= 2\frac{\pi t(1-\lambda)}{r} \int_{2\lambda - \frac{t+r}{t}}^{\frac{t-r}{t}} \zeta^{-1+\mu}(\lambda - \zeta)\, d\zeta \leq C(1-\lambda)\frac{t}{r}\int_{2\lambda - \frac{t+r}{t}}^{\frac{t-r}{t}} \zeta^{-1+\mu} d\zeta \\
&\leq \frac{C(1-\lambda)t}{\mu r}\zeta^{\mu}\Big|_0^{\frac{t-r}{r}} = \frac{C(1-\lambda)t}{\mu r}\left(\frac{t-r}{t}\right)^{\mu}.
\end{aligned}
$$

When $\frac{t+r+1}{2t} \leq \frac{t-r}{t}$, both case above may occur, while only Case 2.2 is possible if the opposite inequality holds true. $\square$

Appendix C

Sup-norm estimate for the Klein-Gordon equation

We provide here a proof of Proposition 3.5.

Lemma C.1 (A decomposition of the Klein-Gordon operator). *For sufficiently smooth solutions v to (3.29), the function $w_{t,x}(\lambda) := \lambda^{3/2}v(\lambda t/s, \lambda x/s)$ is a solution to the second-order ODE in λ*

$$\begin{aligned}&\frac{d^2}{d\lambda^2}w_{t,x}(\lambda) + \frac{c^2}{1+\overline{h}^{00}(\lambda t/s, \lambda x/s)}w_{t,x}(\lambda)\\&= \big(1+\overline{h}^{00}(\lambda t/s, \lambda x/s)\big)^{-1}\big(R_1[v]+R_2[v]+R_3[v]+s^{3/2}f\big)(\lambda t/s, \lambda x/s).\end{aligned}$$

Lemma C.2 (Technical ODE estimate). *Let the function G be defined on some interval $[s_0, s_1]$ and satisfying $\sup|G| \leq 1/3$ and let k be some integrable function defined on $[s_0, s_1]$. The solution z to*

$$\begin{aligned}z''(\lambda) + \frac{c^2}{1+G(\lambda)}z(\lambda) &= k(\lambda),\\ z(s_0) = z_0, \qquad z'(s_0) &= z_1,\end{aligned} \tag{C.1}$$

(for some initial data z_0, z_1) satisfies the uniform estimate for $s \in [s_0, s_1]$

$$\begin{aligned}|z(s)| + |z'(s)| \lesssim& \big(|z_0|+|z_1|+K(s)\big)\\&+\int_{s_0}^{s}\Big(|z_0|+|z_1|+K(\overline{s})\Big)\,|G'(\overline{s})|e^{C\int_{\overline{s}}^{s}|G'(\lambda)|d\lambda}\,d\overline{s}\end{aligned} \tag{C.2}$$

with $K(s) := \int_{s_0}^{s}|k(\overline{s})|\,d\overline{s}$ and a constant $C > 0$.

Proof of Lemma C.1. 1. Flat wave operator. Recall $s = \sqrt{t^2-r^2}$ and $r = |x|$. The flat wave operator $\Box$ in the hyperboloidal frame reads

$$-\Box = \overline{\partial}_0\overline{\partial}_0 - \sum_a \overline{\partial}_a\overline{\partial}_a + 2\sum_a \frac{x^a}{s}\overline{\partial}_0\overline{\partial}_a + \frac{3}{s}\overline{\partial}_0. \tag{C.3}$$

Given any function v, we write

$$w(t,x) = s^{3/2}v(t,x) = (t^2 - |x|^2)^{3/4}v(t,x),$$

and

$$-s^{3/2}\Box v = \overline{\partial}_0\overline{\partial}_0 w - \sum_a \overline{\partial}_a\overline{\partial}_a w + 2\sum_a \frac{x^a}{s}\overline{\partial}_0\overline{\partial}_a w - \frac{3w}{4s^2} - \sum_a \frac{3x^a\overline{\partial}_a w}{s^2}. \tag{C.4}$$

Consider the function of a single variable

$$w_{t,x}(\lambda) := w(\lambda t/s, \lambda x/s) = \lambda^{3/2}v(\lambda t/s, \lambda x/s),$$

so that

$$\frac{d}{d\lambda}w_{t,x}(\lambda) = (\overline{\partial}_0 + s^{-1}x^a\overline{\partial}_a)w(\lambda t/s, \lambda x/s) = \frac{t}{s}\underline{\partial}_\perp w\,(\lambda t/s, \lambda x/s)$$

and

$$\frac{d^2}{d\lambda^2}w_{t,x}(\lambda) = \left(\overline{\partial}_0\overline{\partial}_0 + 2\frac{x^a}{s}\overline{\partial}_0\overline{\partial}_a + \frac{x^a x^b}{s^2}\overline{\partial}_a\overline{\partial}_b\right)w(\lambda t/s, \lambda x/s). \tag{C.5}$$

Combining with (C.4) and recalling $w(t,x) = s^{3/2}v(t,x)$, we obtain

$$\begin{aligned}
&\left(\overline{\partial}_0\overline{\partial}_0 + 2\frac{x^a}{s}\overline{\partial}_0\overline{\partial}_a + \frac{x^a x^b}{s^2}\overline{\partial}_a\overline{\partial}_b\right)w \\
&= -s^{3/2}\Box v + \sum_a \overline{\partial}_a\overline{\partial}_a w + \frac{x^a x^b}{s^2}\overline{\partial}_a\overline{\partial}_b w + \frac{3}{4s^2}w + \sum_a \frac{3x^a}{s^2}\overline{\partial}_a w \qquad (C.6)\\
&= -s^{3/2}\Box v + R_1[v].
\end{aligned}$$

2. Curved wave operator. We write

$$-\Box v = h^{\alpha\beta}\partial_\alpha\partial_\beta v - c^2 v + f$$

and perform a change of frame:

$$\begin{aligned}
h^{\alpha\beta}\partial_\alpha\partial_\beta v =& \overline{h}^{\alpha\beta}\overline{\partial}_\alpha\overline{\partial}_\beta v + h^{\alpha\beta}\partial_\alpha\overline{\Psi}_\beta^{\beta'}\,\overline{\partial}_{\beta'}v \\
=& \overline{h}^{00}\overline{\partial}_0\overline{\partial}_0 v + 2\overline{h}^{0b}\overline{\partial}_0\overline{\partial}_b v + \overline{h}^{ab}\overline{\partial}_a\overline{\partial}_b v + h^{\alpha\beta}\partial_\alpha\overline{\Psi}_\beta^{\beta'}\,\overline{\partial}_{\beta'}v.
\end{aligned}$$

We get

$$\begin{aligned}
-s^{3/2}\Box v =& -s^{3/2}\overline{h}^{00}\overline{\partial}_0\overline{\partial}_0 v - s^{3/2}\big(2\overline{h}^{0b}\overline{\partial}_0\overline{\partial}_b v + \overline{h}^{ab}\overline{\partial}_a\overline{\partial}_b v + h^{\alpha\beta}\partial_\alpha\overline{\Psi}_\beta^{\beta'}\,\overline{\partial}_{\beta'}v\big) \\
&- c^2 s^{3/2}v + s^{3/2}f \\
=& -\overline{h}^{00}\overline{\partial}_0\overline{\partial}_0(s^{3/2}v) - c^2 s^{3/2}v + \overline{h}^{00}\left(\frac{3v}{4s^{1/2}} + 3s^{1/2}\overline{\partial}_0 v\right) \\
&- s^{3/2}\big(2\overline{h}^{0b}\overline{\partial}_0\overline{\partial}_b v + \overline{h}^{ab}\overline{\partial}_a\overline{\partial}_b v + h^{\alpha\beta}\partial_\alpha\overline{\Psi}_\beta^{\beta'}\,\overline{\partial}_{\beta'}v\big) + s^{3/2}f,
\end{aligned}$$

and conclude that

$$-s^{3/2}\Box v = -\overline{h}^{00}\overline{\partial}_0\overline{\partial}_0 w - c^2 w + \overline{h}^{00}\left(\frac{3v}{4s^{1/2}} + 3s^{1/2}\overline{\partial}_0 v\right)$$
$$- s^{3/2}\big(2\overline{h}^{0b}\overline{\partial}_0\overline{\partial}_b v + \overline{h}^{ab}\overline{\partial}_a\overline{\partial}_b v + h^{\alpha\beta}\partial_\alpha \overline{\Psi}_\beta^{\beta'}\,\overline{\partial}_{\beta'} v\big) + s^{3/2} f$$
$$= -\overline{h}^{00}\overline{\partial}_0\overline{\partial}_0 w - c^2 w + R_2[v] + s^{3/2} f. \tag{C.7}$$

Combining (C.6) and (C.7), we get

$$\overline{\partial}_0\overline{\partial}_0 w + 2\frac{x^a}{s}\overline{\partial}_0\overline{\partial}_a w + \frac{x^a x^b}{s^2}\overline{\partial}_a\overline{\partial}_b w - \overline{h}^{00}\overline{\partial}_0\overline{\partial}_0 w + c^2 w = R_1[v] + R_2[v] + s^{3/2} f. \tag{C.8}$$

3. Conclusion. We now write

$$(1+\overline{h}^{00})\left(\overline{\partial}_0\overline{\partial}_0 + 2\frac{x^a}{s}\overline{\partial}_0\overline{\partial}_a + \frac{x^a x^b}{s^2}\overline{\partial}_a\overline{\partial}_b\right) w + c^2 w$$
$$= \overline{h}^{00}\left(2\frac{x^a}{s}\overline{\partial}_0\overline{\partial}_a + \frac{x^a x^b}{s^2}\overline{\partial}_a\overline{\partial}_b\right) w + R_1[v] + R_2[v] + s^{3/2} f$$

and so

$$\left(\overline{\partial}_0\overline{\partial}_0 + 2\frac{x^a}{s}\overline{\partial}_0\overline{\partial}_a + \frac{x^a x^b}{s^2}\overline{\partial}_a\overline{\partial}_b\right) w + \frac{c^2 w}{1+\overline{h}^{00}}$$
$$= (1+\overline{h}^{00})^{-1}\big(R_1[v] + R_2[v] + R_3[v] + s^{3/2} f\big). \tag{C.9}$$

This implies that

$$\frac{d^2}{d\lambda^2} w_{t,x}(\lambda) + \frac{c^2 w_{t,x}(\lambda)}{1+\overline{h}^{00}(\lambda t/s, \lambda x/s)}$$
$$= \big(1+\overline{h}^{00}(\lambda t/s, \lambda x/s)\big)^{-1}\big(R_1[v] + R_2[v] + R_3[v] + s^{3/2} f\big)(\lambda t/s, \lambda x/s). \tag{C.10}$$

□

Proof of Lemma C.2. We consider the vector field $b(\lambda) = \big(z(\lambda), z'(\lambda)\big)^T$ and the matrix $A(\lambda) := \begin{pmatrix} 0 & 1 \\ -c^2(1+G)^{-1} & 0 \end{pmatrix}$ and write $b' = Ab + \begin{pmatrix} 0 \\ k \end{pmatrix}$. Consider the diagonalization $A = PQP^{-1}$ with

$$Q = \begin{pmatrix} ic(1+G)^{-1/2} & 0 \\ 0 & -ic(1+G)^{-1/2} \end{pmatrix}$$

and

$$P = \begin{pmatrix} 1 & 1 \\ \frac{ic}{(1+G)^{1/2}} & -\frac{ic}{(1+G)^{1/2}} \end{pmatrix}, \qquad P^{-1} = \begin{pmatrix} 1/2 & \frac{(1+G)^{1/2}}{2ic} \\ 1/2 & -\frac{(1+G)^{1/2}}{2ic} \end{pmatrix}.$$

We thus have $b' = PQP^{-1}b + \begin{pmatrix} 0 \\ k \end{pmatrix}$, leading us to

$$(P^{-1}b)' = Q(P^{-1}b) + (P^{-1})'b + P^{-1}\begin{pmatrix} 0 \\ k \end{pmatrix}.$$

We regard $(P^{-1})'b$ as a source and write

$$\begin{aligned} P^{-1}b(s) = e^{\int_{s_0}^s Q(\overline{s})d\overline{s}} P^{-1}b(s_0) &+ \int_{s_0}^s e^{\int_\lambda^s Q(\overline{s})d\overline{s}} P^{-1}\begin{pmatrix} 0 \\ k \end{pmatrix} d\lambda \\ &+ \int_{s_0}^s e^{\int_\lambda^s Q(\overline{s})d\overline{s}} (P^{-1})'(\lambda)\, b(\lambda)\, d\lambda. \end{aligned}$$

When $\sup_{\lambda\in[1,s]} |G(\lambda)| \le 1/3$, the norm of $P(\lambda)$ and $P^{-1}(\lambda)$ are bounded for $\lambda \in [s_0, s]$. The norm of $(P^{-1})'(\lambda)$ is bounded by $C|G'(\lambda)|$ for a constant C depending only on c. The norm of Q is bounded by a constant $C > 0$. Observe also that

$$\int_\lambda^s Q(\overline{s})d\overline{s} = \begin{pmatrix} ic\int_\lambda^s (1+G)^{-1/2}(\overline{s})d\overline{s} & 0 \\ 0 & -ic\int_\lambda^s (1+G)^{-1/2}(\overline{s})d\overline{s} \end{pmatrix},$$

therefore

$$e^{\int_\lambda^s Q(\overline{s})d\overline{s}} = \begin{pmatrix} e^{ic\int_\lambda^s (1+G)^{-1/2}(\overline{s})d\overline{s}} & 0 \\ 0 & e^{-ic\int_\lambda^s (1+G)^{-1/2}(\overline{s})d\overline{s}} \end{pmatrix}.$$

The norm of $e^{\int_\lambda^s Q(\overline{s})d\overline{s}}$ is uniformly bounded and we have proven:

$$|z(s)|+|z'(s)| \le C(|z(s_0)|+|z'(s_0)|)+C\,K(s)+C\int_{s_0}^s |G'(\lambda)|(|z(\lambda)|+|z'(\lambda)|)\,d\lambda,$$

and it remains to apply Gronwall's lemma. $\square$

Proof of Proposition 3.5. We have

$$\begin{aligned} w_{t,x}(\lambda) &= \lambda^{3/2}v(\lambda t/s, \lambda x/s), \\ w'_{t,x}(\lambda) &= \frac{3}{2}\lambda^{1/2}v(\lambda t/s, \lambda x/s) + \frac{t}{s}\lambda^{3/2}\underline{\partial}_\perp v(\lambda t/s, \lambda x/s). \end{aligned}$$

The function $w_{t,x}$ is the restriction of $w(t,x) = s^{3/2}v(t,x)$ to the segment $\{(\lambda t/s, \lambda x/s), \lambda \in [s_0, s]\}$. Apply (C.2) and (C.10) to this segment, with

$$s_0 = \begin{cases} 2, & 0 \le r/t \le 3/5, \\ \sqrt{\dfrac{t+r}{t-r}}, & 3/5 \le r/t \le 1. \end{cases}$$

This is the line $\{(\lambda t/s, \lambda x/s)\}$ between (t,x) and the boundary of $\mathcal{K}_{[s_0,+\infty)}$.

The function v is supported in $\mathcal{K}$ and the restriction of v to the hyperboloid $\mathcal{H}_2$ is supported in $\mathcal{H}_2 \cap \mathcal{K}$. We recall that when $3/5 \leq r/t \leq 1$, $w_{t,x}(s_0) = 0$ and when $0 \leq r/t \leq 3/5$, $w_{t,x}(s_0)$ is determined by v_0.

When $0 \leq r/t \leq 3/5$, we apply (C.2) with $s_0 = 2$. When $\lambda = 2$, we write

$$w_{t,x}(2) = w(2t/s, 2x/s) = 2^{3/2} v(2t/s, 2x/s) = 2^{3/2} v_0(2x/s),$$

and

$$\begin{aligned}
w'_{s,x}(2) =& \frac{d}{d\lambda}\big(\lambda^{3/2} v(\lambda t/s, \lambda x/s)\big)\big|_{\lambda=2} \\
=& \frac{3\sqrt{2}}{2} v(2t/s, 2x/s) + 2^{3/2}(s/t)^{-1}\underline{\partial}_{\perp} v(2t/s, 2x/s) \\
=& \frac{3\sqrt{2}}{2} v(2t/s, 2x/s) + 2^{3/2}(s/t)^{-1}\partial_t v(2t/s, 2x/s) \\
&+ 2^{3/2}(x^a/s)\partial_a v(2t/s, 2x/s) \\
=& \frac{3\sqrt{2}}{2} v_0(2x/s) + 2^{3/2}(x^a/s)\partial_a v_0(2x/s) + 2^{3/2}(s/t)^{-1} v_1(2t/s, 2x/s).
\end{aligned}$$

Recall that when $0 \leq r/t \leq 3/5$, we have $4/5 \leq s/t \leq 1$. So we see that $|w_{t,x}(s_0)| + |w'_{t,x}(s_0)| \leq C(\|v_0\|_{L^\infty(\mathcal{H}_2)} + \|v_1\|_{L^\infty(\mathcal{H}_2)})$. Then by (C.2) and (C.10) we have

$$\begin{aligned}
&|w_{t,x}(s)| + |w'_{t,x}(s)| \leq C(\|v_0\|_{L^\infty(\mathcal{H}_2)} + \|v_1\|_{L^\infty(\mathcal{H}_2)}) + CF(s) \\
&+ C(\|v_0\|_{L^\infty(\mathcal{H}_2)} + \|v_1\|_{L^\infty(\mathcal{H}_2)}) \int_2^s |h'_{t,x}(\overline{s})| e^{C\int_{\overline{s}}^s |h'_{t,x}(\lambda)|d\lambda} d\overline{s} \\
&+ C\int_2^s F(\overline{s})|h'_{t,x}(\overline{s})| e^{C\int_{\overline{s}}^s |h'_{t,x}(\lambda)|d\lambda} d\overline{s}.
\end{aligned}$$

When $3/5 \leq r/t \leq 1$, $w_{t,x}(s_0) = w'_{t,x}(s_0) = 0$ and so

$$|w_{t,x}(s)| + |w'_{t,x}(s)| \leq CF(s) + C\int_{s_0}^s F(\overline{s})|h'_{t,x}(\overline{s})| e^{C\int_{\overline{s}}^s |h'_{t,x}(\lambda)|d\lambda} d\overline{s},$$

which leads to $|w_{t,x}(s)| + |w'_{t,x}(s)| \lesssim V(t,x)$. Recall finally $v(t,x) = s^{3/2} w_{t,x}(s)$ and

$$(s/t)^{-1} s^{3/2}\underline{\partial}_{\perp} v(t,x) = w'_{t,x}(s) - \frac{3}{2}s^{1/2} v(t,x) = w'_{t,x}(s) - \frac{3}{2}s^{-1} w_{t,x}(s).$$

□

Appendix D

Commutator estimates for the hyperboloidal frame

In this appendix, we provide some further details on some important properties shared by the commutators arising in our problem. The vector fields ∂_α, and L_a are Killing for the wave operator $\Box$, so that

$$[\partial_\alpha, \Box] = 0, \qquad [L_a, \Box] = 0. \tag{D.1}$$

By introducing

$$[L_a, \partial_\beta] =: \Theta_{a\beta}^\gamma \partial_\gamma, \qquad [\partial_\alpha, \underline{\partial}_\beta] =: t^{-1} \underline{\Gamma}_{\alpha\beta}^\gamma \partial_\gamma, \qquad [L_a, \underline{\partial}_\beta] =: \underline{\Theta}_{a\beta}^\gamma \underline{\partial}_\gamma, \tag{D.2}$$

we find

$$\begin{aligned}
&\Theta_{a0}^\gamma = -\delta_a^\gamma, && \Theta_{ab}^\gamma = -\delta_{ab}\delta_0^\gamma,\\
&\underline{\Gamma}_{0b}^\gamma = -\frac{x^b}{t}\delta_0^\gamma = \Psi_b^0 \delta_0^\gamma, && \underline{\Gamma}_{\alpha 0}^\gamma = 0, \qquad \underline{\Gamma}_{ab}^\gamma = \delta_{ab}\delta_0^\gamma,\\
&\underline{\Theta}_{a0}^\gamma = -\delta_a^\gamma + \frac{x^a}{t}\delta_0^\gamma = -\delta_a^\gamma + \Phi_0^a \delta_0^\gamma, && \underline{\Theta}_{ab}^\gamma = -\frac{x^b}{t}\delta_a^\gamma = \Psi_b^0 \delta_a^\gamma.
\end{aligned} \tag{D.3}$$

All of these coefficients are smooth in the (open) cone $\mathcal{K}$ and homogeneous of degree 0. Furthermore, we also get

$$\underline{\Theta}_{ab}^0 = 0, \qquad \text{so that} \qquad [L_a, \underline{\partial}_b] = \underline{\Theta}_{ab}^c \underline{\partial}_c, \tag{D.4}$$

which means that the commutator of a "good" derivative $\underline{\partial}_b$ with L_a is again a "good" derivative.

Lemma D.1 (Algebraic decomposition of commutators. I). *There exist constants λ_{aJ}^I such that*

$$[\partial^I, L_a] = \sum_{|J| \le |I|} \lambda_{aJ}^I \partial^J. \tag{D.5}$$

Proof. We proceed by induction and, for $|I| = 1$, this is (D.2). Assuming that (D.5) holds for all $|I_1| \le m$, we are going to prove that it is still valid for $|I| \le m+1$. Let $I = (\alpha, \alpha_m, \alpha_{m-1}, \ldots, \alpha_1)$ and $I_1 = (\alpha_m, \alpha_{m-1}, \ldots, \alpha_1)$, so that $\partial^I = \partial_\alpha \partial^{I_1}$. We find

$$\begin{aligned}
[\partial^I, L_a] &= [\partial_\alpha \partial^{I_1}, L_a] = \partial_\alpha([\partial^{I_1}, L_a]) + [\partial_\alpha, L_a]\partial^{I_1} \\
&= \partial_\alpha\Big(\sum_{|J| \le |I_1|} \lambda_{aJ}^{I_1} \partial^J\Big) - \Theta_{a\alpha}^{\gamma} \partial_\gamma \partial^{I_1} \\
&= \sum_{|J| \le |I_1|} \lambda_{aJ}^{I_1} \partial_\alpha \partial^J - \Theta_{a\alpha}^{\gamma} \partial_\gamma \partial^{I_1},
\end{aligned}$$

which yields the statement for $|I| = m+1$. □

Lemma D.2 (Algebraic decomposition of commutators. II). *There exist constants* $\theta_{\alpha J}^{I\gamma}$ *so that*

$$[L^I, \partial_\alpha] = \sum_{|J| \le |I|-1, \gamma} \theta_{\alpha J}^{I\gamma} \partial_\gamma L^J. \tag{D.6}$$

Proof. The case $|I| = 1$ is already covered by (D.2). Assuming that (D.6) is valid for $|I| \le m$, we are going to prove that it is still valid when $|I| = m+1$. We write $L^I = L_a L^{I_1}$ with $|I_1| = m$, and find

$$\begin{aligned}
[L^I, \partial_\alpha] =& [L_a L^{I_1}, \partial_\alpha] = L_a([L^{I_1}, \partial_\alpha]) + [L_a, \partial_\alpha] L^{I_1} \\
=& L_a\Big(\sum_{|J| \le |I_1|-1, \gamma} \theta_{\alpha J}^{I_1\gamma} \partial_\gamma L^J\Big) + \sum_{\gamma} \Theta_{a\alpha}^{\gamma} \partial_\gamma L^{I_1} \\
=& \sum_{|J| \le |I_1|-1, \gamma} \theta_{\alpha J}^{I_1\gamma} L_a \partial_\gamma L^J + \sum_{\gamma} \Theta_{a\alpha}^{\gamma} \partial_\gamma L^{I_1},
\end{aligned}$$

so

$$\begin{aligned}
&[L^I, \partial_\alpha] \\
&= \sum_{|J| \le |I_1|-1, \gamma} \theta_{\alpha J}^{I_1\gamma} \partial_\gamma L_a J^J + \sum_{|J| \le |I_1|-1, \gamma} \theta_{\alpha J}^{I_1\gamma} [L_a, \partial_\gamma] J^J + \sum_{\gamma} \Theta_{a\alpha}^{\gamma} \partial_\gamma L^{I_1} \\
&= \sum_{|J| \le |I_1|-1, \gamma} \theta_{\alpha J}^{I_1\gamma} \partial_\gamma L_a J^J + \sum_{|J| \le |I_1|-1, \gamma} \theta_{\alpha J}^{I_1\gamma} \Theta_{a\gamma}^{\gamma'} \partial_{\gamma'} L^J + \sum_{\gamma} \Theta_{a\alpha}^{\gamma} \partial_\gamma L^{I_1}.
\end{aligned}$$

□

As a consequence of (D.6), we have

$$[\partial^I L^J, \partial_\alpha] u = \sum_{|J'| < |J|, \gamma} \theta_{\alpha J'}^{J\gamma} \partial_\gamma \partial^I L^{J'} u. \tag{D.7}$$

Lemma D.3 (Algebraic decomposition of commutators. III). *One has*

$$[\partial^I L^J, \underline{\partial}_\beta] = \sum_{\substack{|J'|\leq|J|,|I'|\leq|I| \\ |I'|+|J'|<|I|+|J|}} \underline{\theta}^{IJ\gamma}_{\beta I'J'} \partial_\gamma \partial^{I'} L^{J'}, \tag{D.8}$$

where $\underline{\theta}^{IJ\gamma}_{\beta I'J'}$ *are smooth functions satisfying*

$$\left|\partial^{I_1} L^{J_1} \underline{\theta}^{IJ\gamma}_{\beta I'J'}\right| \leq \begin{cases} C\big(|I|,|J|,|I_1|,|J_1|\big)\, t^{-|I_1|} & \text{when } |J'|<|J|, \\ C\big(|I|,|J|,|I_1|,|J_1|\big)\, t^{-|I_1|-1} & \text{when } |I'|<|I|. \end{cases} \tag{D.9}$$

Proof. Consider the identity

$$[\partial^I L^J, \underline{\partial}_\beta] = [\partial^I L^J, \Phi^\gamma_\beta \partial_\gamma] = \Phi^\gamma_\beta [\partial^I L^J, \partial_\gamma] + \sum_{\substack{I_1+I_2=I, J_1+J_2=J \\ |I_1|+|J_1|<|I|+|J|}} \partial^{I_1} L^{J_1} \Phi^\gamma_\beta \partial^{I_2} L^{J_2} \partial_\gamma.$$

Commuting $\partial^{I_2} L^{J_2}$ and ∂_γ, we obtain

$$\begin{aligned}
[\partial^I L^J, \underline{\partial}_\beta] &= \Phi^\gamma_\beta [\partial^I L^J, \partial_\gamma] \\
&\quad + \sum_{\substack{I_1+I_2=I, J_1+J_2=J \\ |I_1|+|J_1|<|I|+|J|}} \partial^{I_1} L^{J_1} \Phi^\gamma_\beta \partial_\gamma \partial^{I_2} L^{J_2} + \sum_{\substack{I_1+I_2=I, J_1+J_2=J \\ |I_1|+|J_1|<|I|+|J|}} \partial^{I_1} L^{J_1} \Phi^\gamma_\beta [\partial^{I_2} L^{J_2}, \partial_\gamma] \\
&= \sum_{\substack{I_1+I_2=I, J_1+J_2=J \\ |I_1|+|J_1|<|I|+|J|}} \partial^{I_1} L^{J_1} \Phi^\gamma_\beta \partial_\gamma \partial^{I_2} L^{J_2} + \sum_{\substack{I_1+I_2=I \\ J_1+J_2=J}} \partial^{I_1} L^{J_1} \Phi^\gamma_\beta [\partial^{I_2} L^{J_2}, \partial_\gamma] \\
&= \sum_{\substack{I_1+I_2=I, J_1+J_2=J \\ |I_1|+|J_1|<|I|+|J|}} \partial^{I_1} L^{J_1} \Phi^\gamma_\beta \partial_\gamma \partial^{I_2} L^{J_2} \\
&\quad + \sum_{\substack{I_1+I_2=I \\ J_1+J_2=J}} \sum_{|J_2'|<|J_2|} \left(\partial^{I_1} L^{J_1} \Phi^\gamma_\beta\right) \theta^{J_2\delta}_{\gamma J_2'} \partial_\delta \partial^{I_2} L^{J_2'}.
\end{aligned}$$

Hence, $\underline{\theta}^{IJ\alpha}_{\gamma I'J'}$ are linear combinations of $\partial^{I_1} L^{J_1} \Phi^\gamma_\beta$ and $\left(\partial^{I_1} L^{J_1} \Phi^\gamma_\beta\right) \theta^{J_2\delta}_{\gamma J_2'}$ and $J_1 + J_2 = J$, which yields (D.8). Note that $\theta^{J_2\delta}_{\gamma J_2'}$ are constants, so that

$$\partial^{I_3} L^{J_3} \left(\partial^{I_1} L^{J_1} \Phi^\gamma_\beta \theta^{J_2\delta}_{\gamma J_2'}\right) = \theta^{J_2\delta}_{\gamma J_2'} \partial^{I_3} L^{J_3} \partial^{I_1} L^{J_1} \Phi^\gamma_\beta.$$

By definition, Φ^γ_β is a homogeneous function of degree zero, so that $\partial^{I_1} L^{J_1} \Phi^\gamma_\beta$ is again homogeneous but with degree ≤ 0. We thus arrive at (D.9). □

Lemma D.4 (Algebraic decomposition of commutators. IV). *One has*

$$[L^I, \underline{\partial}_c] = \sum_{|J|<|I|} \sigma^{Ia}_{cJ} \underline{\partial}_a L^J, \tag{D.10}$$

where the coefficients σ^{Ia}_{cJ} *are smooth functions and satisfy (in* $\mathcal{K}$*)*

$$\left|\partial^{I_1} L^{J_1} \sigma^{Ia}_{cJ}\right| \leq C(|I|,|J|,|I_1|,|J_1|) t^{-|I_1|}. \tag{D.11}$$

Proof. This is also by induction. Again, when $|I| = 1$, (D.10) together with (D.11) are guaranteed by (D.4). Assume that (D.10) and (D.11) hold for $|I| \leq m$, we will prove that they are valid for $|I| = m + 1$. We take $L^I = L_a L^J$ with $|J| = m$, and obtain

$$
\begin{aligned}
[L^I, \underline{\partial}_c] =& [L_a L^J, \underline{\partial}_c] = L_a\big([L^J, \underline{\partial}_c]\big) + [L_a, \underline{\partial}_c] L^J \\
=& L_a\Big(\sum_{|J'|<|J|} \sigma_{cJ'}^{Ja} \underline{\partial}_a L^{J'}\Big) + \underline{\Theta}_{ac}^b \underline{\partial}_b L^J \\
=& \sum_{|J'|<|J|} L_a \sigma_{cJ'}^{Jb} \underline{\partial}_b L^{J'} + \sum_{|J'|<|J|} \sigma_{cJ'}^{Jb} L_a \underline{\partial}_b L^{J'} + \underline{\Theta}_{ac}^b \underline{\partial}_b L^J,
\end{aligned}
$$

so that

$$
\begin{aligned}
[L^I, \underline{\partial}_c] =& \sum_{|J'|<|J|} L_a \sigma_{cJ'}^{Jb} \underline{\partial}_b L^{J'} + \sum_{|J'|<|J|} \sigma_{cJ'}^{Jb} \underline{\partial}_b L_a L^{J'} \\
& + \sum_{|J'|<|J|} \sigma_{cJ'}^{Jb} [L_a, \underline{\partial}_b] L^{J'} + \underline{\Theta}_{ac}^b \underline{\partial}_b L^J \\
=& \sum_{|J'|<|J|} L_a \sigma_{cJ'}^{Jb} \underline{\partial}_b L^{J'} + \sum_{|J'|<|J|} \sigma_{cJ'}^{Jb} \underline{\partial}_b L_a L^{J'} \\
& + \sum_{|J'|<|J|} \sigma_{cJ'}^{Jb} \underline{\Theta}_{ab}^d \underline{\partial}_d L^{J'} + \underline{\Theta}_{ac}^b \underline{\partial}_b L^J.
\end{aligned}
$$

In each term the coefficients are homogeneous of degree 0 (by applying (D.11)), and the desired result is proven. □

The following result is also checked by induction along the same lines as above, and so its proof is omitted.

Lemma D.5 (Algebraic decomposition of commutators. V). *One has*

$$
[\partial^I, \underline{\partial}_c] = t^{-1} \sum_{|J| \leq |I|} \rho_{cJ}^I \partial^J, \tag{D.12}
$$

where ρ_{cJ}^I are smooth functions satisfying

$$
\left|\partial^{I_1} L^{J_1} \rho_{cJ}^I\right| \leq C(|I|, |J|, |I_1|, |J_1|) t^{-|I_1|}. \tag{D.13}
$$

The following statements are now immediate in view of (D.5), (D.6), and (D.10), and (D.12).

Proposition D.1 (Estimates on commutators. I). *For all sufficiently regular functions u defined in the future cone $\mathcal{K}$, one has*

$$\big|[\partial^I L^J, \partial_\alpha]u\big| \leq C(|I|,|J|) \sum_{|J'|<|J|,\beta} |\partial_\beta \partial^I L^{J'} u|, \tag{D.14}$$

$$\big|[\partial^I L^J, \underline{\partial}_c]u\big| \leq C(|I|,|J|)\bigg(\sum_{\substack{|J'|<|J|,a\\|I'|\leq|I|}} |\underline{\partial}_a \partial^{I'} L^{J'} u| + t^{-1} \sum_{\substack{|I|\leq|I'|\\|J|\leq|J'|}} |\partial^{I'} L^{J'} u| \bigg), \tag{D.15}$$

$$\begin{aligned}\big|[\partial^I L^J, \underline{\partial}_\alpha u]\big| \leq & C(|I|,|J|)t^{-1} \sum_{\substack{\beta,|I'|<|I|\\|J'|\leq|J|}} \Big|\partial_\beta \partial^{I'} L^{J'} u\Big| \\ & + C(|I|,|J|) \sum_{\substack{\beta,|I'|\leq|I|\\|J'|<|J|}} \Big|\partial_\beta \partial^{I'} L^{J'} u\Big|,\end{aligned} \tag{D.16}$$

$$\big|[\partial^I L^J, \partial_\alpha \partial_\beta]u\big| \leq C(|I|,|J|) \sum_{\substack{\gamma,\gamma'\\|I|\leq|I'|,|J'|<|I|}} \big|\partial_\gamma \partial_{\gamma'} \partial^{I'} L^{J'} u\big|, \tag{D.17}$$

$$\begin{aligned}&\big|[\partial^I L^J, \underline{\partial}_a \underline{\partial}_\beta]u\big| + \big|[\partial^I L^J, \underline{\partial}_\alpha \underline{\partial}_b]u\big| \\ &\leq C(|I|,|J|)\bigg(\sum_{\substack{c,\gamma,|I'|\leq|I|\\|J'|<|J|}} \big|\underline{\partial}_c \underline{\partial}_\gamma \partial^{I'} L^{J'} u\big| + t^{-1} \sum_{\substack{c,\gamma,|I'|<|I|\\|J'|\leq|J|}} \big|\underline{\partial}_c \underline{\partial}_\gamma \partial^{I'} L^{J'} u\big| \\ &\qquad + t^{-1} \sum_{\substack{\gamma,|I'|\leq|I|\\|J'|\leq|J|}} \big|\partial_\gamma \partial^{I'} L^{J'} u\big| \bigg).\end{aligned} \tag{D.18}$$

Further estimates will be also needed, as now stated.

Proposition D.2 (Estimates on commutators. II). *For all sufficiently regular functions u defined in the future cone $\mathcal{K}$, one has (for all I, J, α)*

$$\big|\partial^I L^J\big((s/t)\partial_\alpha u\big)\big| \leq \big|(s/t)\partial_\alpha \partial^I L^J u\big| + C(|I|,|J|) \sum_{\substack{\beta,|I'|\leq|I|\\|J'|\leq|J|}} \big|(s/t)\partial_\beta \partial^{I'} L^{J'} u\big|. \tag{D.19}$$

Finally, recall from [39] the following technical observation concerning products of first-order linear operators with homogeneous coefficients of order 0 or 1.

Lemma D.6. *For all multi-indices I, the function*

$$\Xi^{I,J} := (t/s)\partial^I L^J (s/t),$$

defined in the closed cone $\overline{\mathcal{K}} = \{|x| \leq t - 1\}$, is smooth and all of its derivatives (of any order) are bounded in $\overline{\mathcal{K}}$. Furthermore, it is homogeneous of degree η with $\eta \leq 0$.

Bibliography

1. M. Ansorg and R.P. Macedo, Spectral decomposition of black-hole perturbations on hyperboloidal slices, Physical Rev. D 93 (2016), 124016.
2. R.A. Asanov, The Schwarzschild metric and de Donder condition, Gen. Relat. Grav. 21 (1989), 149–154.
3. A. Bachelot, Problème de Cauchy global pour des systèmes de Dirac-Klein-Gordon, Ann. Inst. Henri Poincaré 48 (1988), 387–422.
4. A. Bachelot, Asymptotic completeness for the Klein-Gordon equation on the Schwarzschild metric, Ann. Inst. Henri Poincaré: Phys. Théor. 61 (1994), 411–441.
5. F. Beyer and P.G. LeFloch, Dynamics of self–gravitating fluids in Gowdy-symmetric spacetimes near cosmological singularities, Comm. Part. Diff. Equa. (2017). Preprint ArXiv:1512.07187.
6. L. Bieri and N. Zipser, *Extensions of the stability theorem of the Minkowski space in General Relativity,* AMS/IP Studies Adv. Math. 45. Amer. Math. Soc., International Press, Cambridge, 2009.
7. A.Y. Burtscher and R. Donninger, Hyperboloidal Evolution and Global Dynamics for the Focusing Cubic Wave Equation, Comm. Math. Phys. 353 (2017), 549–596.
8. A.Y. Burtscher and P.G. LeFloch, The formation of trapped surfaces in spherically-symmetric Einstein-Euler spacetimes with bounded variation, J. Math. Pures Appl. 102 (2014), 1164–1217.
9. Y. Choquet-Bruhat, *General relativity and the Einstein equations,* Oxford Math. Monograph, Oxford Univ. Press, 2009.
10. D. Christodoulou, Global solutions of nonlinear hyperbolic equations for small initial data, Comm. Pure Appl. Math. 39 (1986), 267–282.
11. D. Christodoulou, *The formation of black holes in general relativity,* Eur. Math. Soc. (EMS) series, Zürich, 2008.
12. D. Christodoulou and S. Klainerman, *The global nonlinear stability of the Minkowski space,* Princeton Math. Ser. 41, Princeton University, 1993.
13. P.T. Chrusciel, Anti-gravité à la Carlotto-Schoen, Bourbaki seminar, Paris, November 2016, to appear.

14. P.T. CHRUSCIEL AND E. DELAY, Exotic hyperbolic gluings, Jour. Differ. Geom. (2017). Preprint ArXiv:1511.07858.
15. J. CORVINO AND R. SCHOEN, On the asymptotics for the vacuum Einstein constraint equations, J. Diff. Geom. 73 (2006), 185–217.
16. J.-M. DELORT, Existence globale et comportement asymptotique pour l'équation de Klein-Gordon quasi-linéaire à données petites en dimension 1, Ann. Sci. École Norm. Sup. 34 (2001), 1–61.
17. J.-M. DELORT, D. FANG, AND R. XUE, Global existence of small solutions for quadratic quasilinear Klein-Gordon systems in two space dimensions, J. Funct. Anal. 211 (2004), 288–323.
18. D. FAJMAN, J. JOUDIOUX, AND J. SMULEVICI, A vector field method for relativistic transport equations with applications, Analysis & PDE (2017), to appear. See also ArXiv:1510.04939.
19. D. FAJMAN, J. JOUDIOUX, AND J. SMULEVICI, Sharp asymptotics for small data solutions of the Vlasov-Nordström system in three dimensions, Preprint ArXiv:1704.05353.
20. J. FRAUENDIENER, Numerical treatment of the hyperboloidal initial value problem for the vacuum Einstein equations. II. The evolution equations, Phys. Rev. D 58 (1998), 064003.
21. J. FRAUENDIENER AND M. HEIN, Numerical evolution of axisymmetric, isolated systems in general relativity, Phys. Rev. D 66 (2002), 124004.
22. H. FRIEDRICH, On the regular and the asymptotic characteristic initial value problem for Einstein's vacuum field equations, Proc. R. Soc. London Ser. A 375 (1981), 169–184.
23. H. FRIEDRICH, Cauchy problems for the conformal vacuum field equations in General Relativity, Commun. Math. Phys. 91 (1983), 445–472.
24. P. GERMAIN, Global existence for coupled Klein–Gordon equations with different speeds, Ann. Inst. Fourier 61 (2011), 2463–2506.
25. N. GRUBIC AND P.G. LEFLOCH, On the area of the symmetry orbits in weakly regular Einstein-Euler spacetimes with Gowdy symmetry, SIAM J. Math. Anal. 47 (2015), 669–683.
26. D. HILDITCH, E. HARMS, M. BUGNER, H. RUETER, AND B. BRUEGMANN, The evolution of hyperboloidal data with the dual foliation formalism: Mathematical analysis and wave equation tests, Preprint ArXiv:1609.08949.
27. L. HÖRMANDER, *Lectures on nonlinear hyperbolic differential equations,* Springer Verlag, Berlin, 1997.
28. A. HOSHIGA AND H. KUBO, Global small amplitude solutions of nonlinear hyperbolic systems with a critical exponent under the null condition, SIAM J. Math. Anal. 31 (2000), 486–513.
29. A.D. IONESCU AND B. PAUSADER, Global solutions of quasilinear systems of Klein-Gordon equations in 3D, J. Eur. Math. Soc. 16, 2355–2431.
30. S. KATAYAMA, Global existence for coupled systems of nonlinear wave and Klein-Gordon equations in three space dimensions, Math. Z. 270 (2012), 487–513.
31. S. KATAYAMA, Asymptotic pointwise behavior for systems of semilinear wave equations in three space dimensions, J. Hyperbolic Differ. Equ. 9 (2012),

263–323.
32. S. Klainerman, Global existence for nonlinear wave equations, Comm. Pure Appl. Math. 33 (1980), 43–101.
33. S. Klainerman, Global existence of small amplitude solutions to nonlinear Klein-Gordon equations in four spacetime dimensions, Comm. Pure Appl. Math. 38 (1985), 631–641.
34. S. Klainerman, The null condition and global existence to nonlinear wave equations, in "Nonlinear systems of partial differential equations in applied mathematics", Part 1, Santa Fe, N.M., 1984, Lectures in Appl. Math., Vol. 23, Amer. Math. Soc., Providence, RI, 1986, pp. 293–326.
35. S. Klainerman, Remarks on the global Sobolev inequalities in the Minkowski space $\mathbb{R}^{n+1}$, Comm. Pure Appl. Math. 40 (1987), 111–117.
36. S. Klainerman and I. Rodnianski, On the formation of trapped surfaces, Acta Math. 208 (2012), 211–333.
37. S. Klainerman, I. Rodnianski, and J. Szeftel, The bounded L2 curvature conjecture, Invent. Math. 202 (2015), 91–216.
38. P.G. LeFloch, An introduction to self-gravitating matter, Graduate course given at Institute Henri Poincaré, Paris, Fall 2015, available at *http://www.youtube.com/user/PoincareInstitute.*
39. P.G. LeFloch and Y. Ma, *The hyperboloidal foliation method,* World Scientific, 2014.
40. P.G. LeFloch and Y. Ma, The global nonlinear stability of Minkowski spacetime for the Einstein equations in presence of massive fields, Note C.R. Acad. Sc. Paris 354 (2016), 948–953.
41. P.G. LeFloch and Y. Ma, The global nonlinear stability of Minkowski space for self-gravitating massive fields. The wave-Klein-Gordon model, Comm. Math. Phys. 346 (2016), 603–665.
42. P.G. LeFloch and Y. Ma, The global nonlinear stability of Minkowski space for the $f(R)$-theory of modified gravity, in preparation.
43. P.G. LeFloch and A.D. Rendall, A global foliation of Einstein-Euler spacetimes with Gowdy-symmetry on T^3, Arch. Rational Mech. Anal. 201 (2011), 841–870.
44. H. Lindblad, M. Nakamura, and C.D. Sogge, Remarks on global solutions for nonlinear wave equations under the standard null conditions, J. Differential Equations 254 (2013), 1396–1436.
45. H. Lindblad and I. Rodnianski, Global existence for the Einstein vacuum equations in wave coordinates, Comm. Math. Phys. 256 (2005), 43–110.
46. H. Lindblad and I. Rodnianski, The global stability of Minkowski spacetime in harmonic gauge, Ann. of Math. 171 (2010), 1401–1477.
47. Y. Ma, A quasi-linear wave-Klein-Gordon system in $2+1$ dimensions, in preparation.
48. V. Moncrief and O. Rinne, Regularity of the Einstein equations at future null infinity, Class. Quant. Grav. 26 (2009), 125010.
49. H. Okawa, V. Cardoso, and P. Pani, Collapse of self-interacting fields in asymptotically flat spacetimes: do self-interactions render Minkowski spacetime unstable?, Phys. Rev. D 89 (2014), 041502.

50. O. RINNE, An axisymmetric evolution code for the Einstein equations on hyperboloidal slices, Class. Quantum Grav. 27 (2010), 035014.
51. J. SHATAH, Global existence of small solutions to nonlinear evolution equations, J. Differential Equations 46 (1982), 409–425.
52. J. SHATAH, Normal forms and quadratic nonlinear Klein–Gordon equations, Comm. Pure Appl. Math. 38 (1985), 685–696.
53. J. SMULEVICI, Small data solutions of the Vlasov-Poisson system and the vector field method, Preprint ArXiv:1504.02195.
54. J. SPECK, The global stability of the Minkowski spacetime solution to the Einstein-nonlinear system in wave coordinates, Analysis & PDE 7 (2014), 771–901.
55. W.A. STRAUSS, *Nonlinear wave equations,* CBMS 73, Amer. Math. Soc., Providence, 1989.
56. A. VANÓ-VINUALES, S. HUSA, AND D. HILDITCH, Spherical symmetry as a test case for unconstrained hyperboloidal evolution, Class. Quantum Gravity 32 (2015), 175010.
57. A. ZENGINOGLU, Hyperboloidal evolution with the Einstein equations, Class. Quantum Grav. 25 (2008), 195025.
58. A. ZENGINOGLU, Hyperboloidal layers for hyperbolic equations on unbounded domains, J. Comput. Phys. 230 (2011), 2286–2302.

Printed in the United States
By Bookmasters